T0258081

IEE HISTORY OF TECHNOLOGY SERIES 18

Series Editor: Dr B. Bowers

LORD KELVIN

his influence on electrical measurements and units

Other volumes in this series:

LORD KELVIN

his influence on electrical measurements and units

PAUL TUNBRIDGE

Peter Peregrinus Ltd. on behalf of the Institution of Electrical Engineers

Published by: Peter Peregrinus Ltd., London, United Kingdom

Peter Peregrinus Ltd.,
Michael Faraday House,
Six Hills Way, Stevenage,
Herts. SG1 2AY, United Kingdom

British Library Cataloguing in Publication Data

A CIP catalogue record for this book
is available from the British Library.

ISBN 0 86341 237 8

Printed in England by Bookcraft, Avon

Contents

Acknowledgements

My thanks go to Mr John T.Lloyd, B.Sc, of Glasgow University, who as Keeper of the Kelvin Museum, kindly provided me with copies of his lecture to the IEE (1982) and article 'Lord Kelvin and his measuring instruments', from which the histogram of Kelvin's personal history has been reproduced (*Electronics and Power*, 10.1.1974). I would thank also Dr Brian Bowers for his valuable advice and for granting me access to the extensive Crompton Collection of documents preserved in the Science Museum Kensington.

A number of Kelvin's unpublished letters or drafts, originate from my visits to the Kelvin Collection of Cambridge University, the Public Record Office, Glasgow University Library, Public and University Library of Geneva, International Electrotechnical Commission, Geneva, House of Lords Records Office, National Physical Laboratory, British Association for the Advancement of Science, The Cromptonian Association, Association Suisse des Electriciens, the Archives of the Institution of Electrical Engineers and of the Institution of Civil Engineers, Institution of Mechanical Engineers, Science Museum Library, Imperial College of Science & Technology Archives, and the Royal Institution, all of whose assistance and advice has proved most helpful.

In acknowledging their permission to reproduce unpublished letters and documents, I also wish to express my appreciation to the Editors of *Notes and Records of the Royal Society London*, and of *Multilingua*, for the reproduction in part of my printed articles on Lord Kelvin and the International Electrotechical Vocabulary, respectively.

Introduction

The history of scientific units and standards is directly linked with Lord Kelvin, whose name has been assigned by the Bureau International des Poids et Mesures to the unit of thermodynamic temperature as one of the basic units of the *Système International d'Unités*.

Kelvin, who is recognised as Britain's most outstanding scientist after Newton, was the author of many textbooks and hundreds of scientific papers. Most of his prolific correspondence is preserved in the Kelvin Collection of Cambridge University Library. A major part of Kelvin's work throughout his life was devoted to the development of electrical units and standards which, in an historical context, can be conveniently considered under three distinct phases, with all of which Kelvin was intimately involved:

- *the discovery of fundamental concepts, and subsequent codification of this knowledge into universally understood laws.* By the middle of the 19th century this had progressed to the point where little remained to be uncovered. Kelvin's mathematically trained brain and creative thinking enabled him to seek out and develop these basic principles into measured scientific and electrotechnical applications;

- *international adoption for scientific purposes of the metric system* (the standard of standards as it has been described, which by 1875 was recognized in most civilized nations), and the creation of uniform electrical units and standards based on the work of the British Association for the Advancement of Science which continued under Kelvin's inspired leadership from 1860 into the 20th century;

- *the great International Electrical Congress held in Paris in 1881*, at which Kelvin was a key figure in securing the first worldwide adoption of electrical units and standards. The unprecedented growth in the electrical industry led to a series of international congresses and conferences. In 1904, the Sixth International Electrical Congress held in the United States recognized the need for the creation of two permanent international commissions entrusted respectively to consider:

  (i) the uniformity of electrical units and standards,
  (ii) the standardization of the nomenclature and ratings of electrical apparatus and machinery.

The first Commission met in London, where an International Conference was held in 1908 on electrical units and their physical representation. The second, the International Electrotechnical Commission (IEC) came into being in 1906 with Lord Kelvin as the first elected President and Colonel R.E.Crompton as Honorary Secretary.*

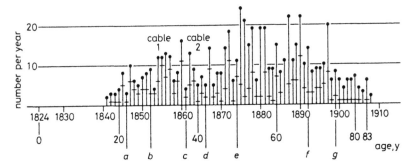

Fig. 1   Histogram of Kelvin's publications and events in his personal history.

   a) Created Professor at Glasgow University
   b) Kelvin's first marriage
   c) Accident leaving Kelvin with permanent limp
   d) Knighthood for his work on cables
   e) Kelvin's second marriage
   f) Created a peer
   g) Retired from Glasgow University Professorship

   Publications: height of bar through vertical lines shows number of electrical and magnetic publications

* The IEC is briefly referred to in Chapters 6, 7 and 8, and Appendix D. Based in Geneva, its role today extends far beyond nomenclature and ratings. The IEC, now composed of 41 National Committees, in close collaboration with other international organizations and in particular the ISO (International Organization for Standardization), prepares standards developed by its more than 200 specialised committees and sub-committees. Covering most electrotechnical fields, the work still includes such fundamental subjects as terminology, quantities and units, symbols, standard voltages, current ratings and frequencies, measuring instruments, meters, etc.

Chapter 1
Kelvin and electrical science

Kelvin's name is to be found in almost every sector of 19th century science. The abstruse nature of much of Kelvin's advanced work renders it obscure to the non-scientific reader. This factor as well as the extent of his papers, most of which are preserved in the Kelvin Collection of Cambridge University Library, explains why no definitive biography has superseded Silvanus P. Thompson's two-volume life published in 1910, three years after Kelvin's death.

William Thomson, the son of a professor of mathematics was born in Belfast on June 26, 1824. He was knighted for his work on the Atlantic telegraph in 1866, was President of the Royal Society from 1890 to 1894, three times President of the Institution of Electrical Engineers, and was raised to the peerage in 1892 when he became Lord Kelvin.

This change of name led to some confusion, and on one occasion a distinguished foreign engineer asked who this person Kelvin was, 'who was claiming to have invented the galvanometer that the whole world knows to have been invented by Sir William Thomson'.[1]

When he was eight years old the family had moved to Glasgow where his father took up a post at the University. Directly following his matriculation, at the age of ten William entered Glasgow University.

During his fifth year, he read in two weeks Fourier's *Théorie analytique de la chaleur*, whose content was to influence his whole career. This prompted him to write his first paper justifying Fourier's method against the criticisms made by Professor Kelland of Edinburgh University who, in his 'Theory of Heat' (1837), had written that 'nearly all M. Fourier's Series in this branch of the subject are erroneous.' Thomson's father sent a copy of the paper to Kelland, which 'after being toned down somewhat, was published'.

His fluency in French and German enabled him to follow all the scientific developments in Europe, and later to take a lead in international standards activities where he was outstanding in all aspects of scientific committee work.

He left Glasgow in 1841 – without bothering to take a degree – on entry to Cambridge University. As an undergraduate of seventeen, his third published paper 'The uniform motion of heat in homogenous solid bodies, and its connection with the mathematical theory of electricity' appeared in the *Cambridge Mathematical Journal* in 1842. Until 1845 he was mainly concerned with the application of mathematical analysis to physics.

He was appointed professor of natural philosophy in Glasgow University at the age of twenty-two. By this time, he had already published more than twenty scientific papers, four of which were in French. His paper 'On the mathematical theory of electricity in equilibrium', published in the *Cambridge and Dublin Mathematical Journal* in 1845 was of sufficient importance to be reprinted ten years later.[2]

Early in 1845, Thomson made the acquaintance of Michael Faraday for whom he retained the greatest admiration. One of his most treasured possessions, and which he would proudly show to his students, was a piece of heavy glass used by Faraday to demonstrate the connection between light and electricity.

On the basis of the often intuitive and inspired discoveries made by the early 19th century scientists, Thomson's analytical and mathematically trained brain enabled him to penetrate even deeper the as yet unexplored sciences including electricity and heat.[3]

In June 1847, he communicated to the Cambridge Philosophical Society a paper 'On an absolute thermometric scale founded on Carnot's theory of the motive power of heat, and calculated from Regnault's observations'.

He demonstrated that 'such a scale is obtained in terms of Carnot's theory, each degree being determined by the performance of equal equations of work in letting one unit of heat be transformed in being let down through that difference of temperature. This indicates as the absolute zero of temperature the point which would be marked as $-273°$ on the air-thermometer scale'.[4]

The subsequent researches of Joule and Thomson on the 'Joule-Kelvin effect' led to the 'thermodynamic scale' (ie. the thermometric standard of temperature measure) proposed by Thomson, which became the basis of the Kelvin scale. The *kelvin* is the name now given to the unit of thermodynamic temperature, one of the units of the International System of Units (S.I.).[5]

In the years 1851 to 1854, Thomson formulated the two great laws of thermodynamics: the law of equivalence discovered by Joule, and the law of transformation which he generously attributed to Carnot and Clausius. In 1855 he introduced the concept of 'available energy' which was the foundation of later developments in thermodynamics.[6]

In 1849, he arrived at a complete solution of the problem of the mutual attraction between two electrified spherical conductors. Published in the *Philosophical Magazine* in 1853, this paper demonstrated his skill in applying mathematical methods to resolve difficult physical problems.[7]

Thomson became famous as a result of his work on the Atlantic Cable when, following the failure in 1857, he was called upon to assist more actively in the work. Throughout the trials, the incident of 1865 when 1000 miles of cable were lost, and the success in 1866 when the new cable was laid and the lost one recovered from the ocean and completed, Sir William Thomson – who was knighted for his achievements – was the ruling spirit behind the work.[8]

In 1872, Thomson published his reprint of papers on electricity and magnetism. In 1879, together with P.G.Tait of Edinburgh University, a new edition of their *Natural Philosophy* was published. The first sentence of

Fourier's *Theory of heat* is reproduced in the original French at the top of their preface:

> 'Primary causes are unknown to us; but they are subject to simple and constant laws, which may be discovered by observation, and the study of which is the object of natural philosophy.'

The invention by Faure in 1881 of his accumulator stimulated Thomson to write to *The Times* and *Nature* on the importance of this invention. That same year he collaborated with Ferranti in designing a special winding for an alternating current dynamo. It was from about this date that Thomson began to take a greater interest in the generating problems of electrical engineers.[9]

In this way, Thomson acted as a consultant to Ferranti and other leading engineers such as Crompton, with whom he had a long and productive association. On his being elected in 1906 the first President of the International Electrotechnical Commission (IEC), Col. Crompton on the same occasion was appointed Honorary Secretary.

One of his closest friends, and as he put it 'a loved and highly esteemed colleague', was Prof Eleuthère Mascart (1837–1908). In 1896 on the occasion of the Jubilee of Thomson (who by now had been raised to the peerage as Lord Kelvin) at Glasgow University, Mascart presented him with the gold Arago Medal awarded by the Académie des Sciences.

Mascart was so overcome with emotion on this memorable occasion that he was obliged to hand his unread speech to Lady Kelvin. He mentioned Kelvin's expressed gratitude to such inspired mentors as Fourier, Laplace and Carnot, and to France as having been the *alma mater* of his scientific youth. He felt that Kelvin had more than repaid his debt by having successfully tackled all the questions of natural philosophy by his contributions to progress in scientific theories and in their practical implementation in the development of industry.[10, 11]

Although perhaps less exciting, Kelvin's greatest contribution to science was undoubtedly his work – which continued throughout the second half of the 19th century – on the establishment of international electrical units and standards. His leadership, recognised throughout the world in this vital sector, and in particular of the British Association for the Advancement of Science, resulted in the laying of solid foundations for subsequent developments in electrical science.

The importance of Kelvin's work on electrical standards was emphasised at the First International Electrical Congress held in Paris, which opened up a new era in scientific progress. On the final day of the Congress, October 5, 1881, M. J-B Dumas, Permanent Secretary of the Académie des Sciences, devoted his closing speech solely to the achievements of the British Association in this field.

He referred to the birth of a vast industry founded on electrical power, coupled with all the possibilities of the new form of communication which now traversed oceans, vast distances and every other obstacle. He emphasised, however, that such progress had been inconsistent with the diversity of measurements employed in the different countries to designate this force.

The installation of telegraphic apparatus linking different countries had involved long and useless calculations before complete agreement had been obtained. It was not just a question of each nation, but each engineer, whose sole aim it seemed had been to introduce new electrical units of measurement. This trend had been on the increase when the British Association decided to intervene. Basing their work on the discoveries of those early 19th century scientists, whose names would outlive all those celebrated in politics and other fields, the British Association had succeeded after years of effort in establishing a closely coordinated system of electrical measurements.

Whether it was a question of mechanical or electrical force, or electricity related to heat generation or to chemical decomposition, henceforth, all this would be related to a common measurement derived from three absolute units which could be formulated in clear and precise terms without any risk of misunderstanding.

It was in the light of such a scientific monument, worthy of the greatest respect, that the Congress had accomplished the task before it. While the Congress had not hesitated to adopt the principles laid before them by the British Association, the distinguished British delegates for their part, had readily accepted those changes in detail rendered necessary by the state of the art, which had facilitated the universal adoption of the system.[12, 13]

Many years later, at a banquet held on the occasion of the Kelvin Centenary in London on July 11, 1924, Sir Richard Glazebrook, FRS, a Past President of the Institution of Electrical Engineers, and first Director of the National Physical Laboratory, proposed the toast in honour of Lord Kelvin. He reminded the distinguished gathering of scientists that the foundations of nearly all they had learnt had rested on the work of Kelvin. In asking those present to imagine a world in which Kelvin had never lived or worked, he gave pride of place to his work on electrical units:

> 'Try to think of it without the C.G.S. system of units, without our knowledge of the importance and value of the second law of thermo-dynamics, without any real information as to what goes on when an alternating electric current circles round a wire – producing as we now know, all the phenomena of wireless telegraphy – without the Atlantic cable, the mirror galvanometer, without the compass or the deep-sea sounding gear. All those we owe to Lord Kelvin.'[14]

Notes

(1)   Silvanus P.Thompson, *The Life of William Thomson*, Baron Kelvin of Largs, 2 vols, London, 1910,vol II,pp 909,913.
Silvanus P.Thompson, *Lord Kelvin*, published by the International Electrotechnical Commission (IEC), Fascicule 13, London, 1913, p 26
Alexander Russell, *Lord Kelvin: his Life and Work*, London, 1912, Preface.

(2) NOTE
By the time of his death in 1907, Lord Kelvin had published 25 textbooks including a three-volume book on mathematics, 661 scientific papers, and 70 patents – most of which were successful inventions and, what is more unusual, were financially profitable. He had been elected to nearly 90 learned societies and academies throughout the world (including the Societé des Arts; and Honorary Membership of the Société de physique et d'histoire naturelle de Genève, which, following his death in 1907, published a three-page obituary notice in the *Rapport du Président* for that year).

(3) NOTE
His first duty as professor in November 1846 was to read a long 'Introductory lecture to the course on natural philosophy'. He emphasized to the students that 'the first work is to observe and classify facts; the process of inductive generalisation follows, in which the laws of nature are the objects of research..' Of heat, electricity and magnetism, he said, 'Our knowledge of these branches of the science is not so far advanced as to enable us to reduce all the various phenomena to a few simple laws from which, as in mechanics, by means of mathematical reasoning every particular result may be obtained; but observation and experiment are the principal means by which our knowledge in this department can be enlarged'.
In scientific research, he urged them when tackling a difficult problem or trying to understand a new principle to 'have a share of that spirit of enterprise which led Newton on to his investigations; and when the problem is solved, when the doubts have vanished, a feeling of satisfaction will be the reward, similar to that which Newton must have felt after some of his great discoveries'. He insisted on the importance of the principles of natural philosophy as opposed to their practical applications which might or might not result from discoveries. He gave the example of Oersted who 'would never have made his great discovery of the action of galvanic currents on magnets had he stopped in his researches to consider in what manner they could possibly be turned to practical account...'
(*Life*, op cit, pp 190,239,245,247,249)

(4) *Lord Kelvin*, (IEC) op cit, p 10

(5) Paul Tunbridge,A letter by William Thomson,FRS, on the 'Thomson Effect', *Notes and Records of the Royal Society*, Vol 26, No 2, Dec.1972, pp 229–232 (See App. E)
Van Nostrand's Scientific Encyclopedia,3rd ed, 1958, p 1680.
BIPM, *Le Système International d'Unités*,Breteuil,1981, pp 8,33,34

(6) *Lord Kelvin*, (IEC), op cit, p 12

(7) On the mutual attraction and repulsion between two electrified spherical conductors, *Phil Mag*.V, April 1853, pp 287–297

(8) *Lord Kelvin* (IEC), op cit, p16

(9) On some uses of Faure's accumulator in connection with lighting by electricity, *B.A.Report*, 1881, p 526
NOTE
Thomson's letter of June 6 in *The Times* praising Faure's accumulator was criticised for having 'trop de fleurs, pas assez de chiffres' by *L'Electricien* (1.7.1881,pp 270–276)

(10) *Life*, op cit, pp 978,979

(11)   NOTE
Mascart's career as a scientist was a distinguished one; after serving as permanent Secretary, in 1904 he was to be elected President of the Académie des sciences. Like Thomson, Mascart was an outstanding mathematician and experimental physicist who, in addition to being a foreign member of the Royal Society, was the first foreigner to be selected in 1900 as Vice-President of the Institution of Electrical Engineers. He had been the assistant of Régnault, succeeding him as professor of physics in 1872. He thus had much in common with Thomson who had also in 1845 worked in the laboratory of Régnault 'by whom he had been taught a faultless technique, a love of precision in all things, and the highest virtue of the experimenter – patience'. (a) (b)
In 1882, Thomson received a copy of Mascart's first volume of *Lécons sur l'électricité et magnétisme*, which in part dealt with some of Kelvin's work. Thomson in acknowledging receipt of the book, shows how they both shared a similar approach to science in their respect for a balanced combination of theory and practice:

'. . . Je le salue comme un organe de haute valeur pour l'expansion de nos connaissances scientifiques sur ce sujet, dans une forme qui tend à la fois à l'augmentation de nos connaissances par les recherches récentes et à l'intelligence des applications pratiques de ce qui est déjà connu.'

Had Mascart lived, he would as the president-elect (and first President of the French Electrotechnical Committee of the IEC) have succeeded Kelvin as President of the IEC. (c)

(a) *DSB*
(b) *Life*, op cit, pp 1125,1126,1154
(c) P.Janet, *La vie et les oeuvres de E.Mascart,* Commission Electrotechnique Internationale, Comité Electrotechnique Français, Paris, 1910, IEC Fasicule No 2, pp 24, 44
Report of IEC Council, Oct 1908,pp 22,23

(12)   Louis Figuier, *Les nouvelles conquêtes de la science*, Paris 1887, pp 594,595 (French original).

(13)   NOTE
Members of the B.A.Committee on standards for electrical measurements. (*indicates Members of the committee in 1912 listed in *Reports of the Committee on Electrical Standards*, London,1913):

1862–70	1881–1907	Lord Kelvin	

1862–70	Prof A.Williamson		
1867–70	Mr D.Forbes	1867–70 1881	Mr Hockin
1862–70	Sir Charles Wheatstone	1881–1908	Prof W.Ayrton
1862–70	Prof W.H.Miller	1881*	Prof J.Perry
1862–70	Dr A.Matthiessen	1881*	Prof W.Adams
1862–70	1881–1884 Prof Fleeming Jenkin	1881*	Lord Rayleigh
1863–70	Mr C.F.Varley	1881*	Sir Oliver Lodge
1863–70	Prof Balfour Stewart	1881–97	Dr Hopkinson
1863–70	Mr C.W.Siemens ·	1881*	Dr Muirhead
1863–70	Prof J.Clerk Maxwell	1881*	Sir W.Preece
1863–70	Dr Joule	1881–1897	Mr H.Taylor
1863–70	Sir Charles Bright	1882–1904	Prof Everett

1863	Dr Esselbach	1882*	Prof Schuster
1867–70	1881* Prof G.C.Foster	1883	Sir W.Siemens
1867–70	Mr Latimer Clark	1883*	Dr J.Fleming

(14) *Kelvin Centenary and Addresses,* London,1924, p 42
  NOTE
  Among the addresses presented on this occasion, the President of the
  International Electrotechnical Commission, M.Guido Semenza (Italy), on behalf
  of his 'colleagues of all nations expressed their profound gratitude for the
  advice and support which Lord Kelvin, as its first President in 1906, gave to this
  world-wide movement in its initial stages, and which has played so great a part
  in whatever success the Commission has so far attained'.*(Kelvin Centenary,* op cit,
  p 93)

Chapter 2
Kelvin and the metric system

Measuring systems were in use by the Greeks, and before that by the Egyptians whose pyramids could not have been designed and constructed without reliable measuring systems. By the time of the Roman Empire, the measuring standards were carefully preserved in the temples and emperors' palaces to safeguard their integrity. With the disintegration of the Roman Empire the uniform system of measurement fell into disuse.[1]

In the 17th and 18th centuries it became imperative for the various measuring systems employed in Europe to be brought into line not only to enable merchants to sell their manufactured goods but to permit scientists of various countries to compare their experiments and evaluate their findings.[2]

No real improvement became possible until the adoption in France of metric units. The history of the metric system is a long and complicated one but it is generally recognized that it was Gabriel Mouton, a clergyman in Lyon, who first proposed in 1670 a system of linear measures based on decimal divisions.[3]

A century later, in England, James Watt writing to a friend in 1783, complained of the difficulties of comparing scientific results from different countries. He proposed they should agitate for the adoption of an international unit of weights and measures for scientific purposes. He wrote to French savants on the subject which triggered agitation in France.

As a result, in 1790, and despite the revolutionary events in France, Talleyrand brought in a bill before the Legislative Assembly proposing the setting up of a Commission. It was a provision of that measure that The Royal Society of London and the French Academy should nominate members of the Commission since it was considered that the matter should be dealt with at international level. It is a matter of regret that France and England were at war at the time and The Royal Society declined to accept the invitation.[4]

A committee of French scientists, including Lagrange and Laplace, without the assistance of the English, reported their conclusions in March 1791. The choice of a fundamental unit as the basis of a rational system of measures was to be decided taking a terrestial arc of meridian between Dunkirk and Barcelona. From the distance thus determined the length of the entire quadrantal arc from the pole to the equator was to be calculated.

The ten-millionth part of the total calculated length was then to be taken as the base or fundamental unit of length, the *mètre* (from the Greek *metron*, a measure), and precisely marked off on a number of metal bars for use as working standards in science and commerce. The length of the unit was provisionally fixed at 3 *pieds* 11.44 *lignes* based on calculations made of a meridian in France by Lacaille in 1740.

The geodetic survey work to measure the earth's meridian quadrant at sea level between Dunkirk and Barcelona, begun in 1792, was often disturbed by harassment from unruly elements of the population in the more inaccessible regions. In April 1796, therefore, two years before the *mètre* had been calculated from the geodetic survey, the French Government established the metric system based on the provisional standard together with the nomenclature of the metric units introduced.

In spite of the non-participation of the British and United States (whose Government under Washington in 1795 had also rejected the metric system), the French government with remarkable foresight and courage went ahead, so that by June 1799 a platinum prototype *mètre* and *kilogrammme* had been constructed with iron copies for wider distribution.[5]

2.1 The Metric Convention

The development of scientific applications in Europe and the rest of the world called for a new appraisal of the accuracy of the standard mètre. Under the auspices of the French Government an international conference held in 1872 attended by scientists representing 26 countries agreed on the construction of new practical *mètre* and *kilogramme* standards of a new metal, but based on the exact dimensions of the units made 75 years previously and preserved in Paris.

In 1875, the *Convention Mètre* was signed in Paris by 18 countries including the USA and Russia, but with Britain and Holland abstaining. The French government offered the Pavilion de Breteuil as a repository for the new standards which it designated the Bureau International des Poids et Mésures (International Bureau of Weights and Measures).

This new organization, the BIPM, was entrusted with custody of the international metric prototypes and official comparisons with national standards, comparison of other units with the metric standards, and other scientific standardization work related to international metrology.[6]

The BIPM continued to influence the development of international metrological standards, and in 1921 the Sixth General Conference on Weights and Measures decided to revise the Metre Convention by extending its responsibility to include electrical and photometric units. Some twenty-seven years later on January 1, 1948, the absolute electrical units (defined as integral powers of ten of corresponding units in the electromagnetic CGS system) replaced the international units.[7, 8]

Notes

(1)   H.Moreau, *Le système métrique*, Paris, 1975, pp 3–5
       Rexmond C.Cochrane, U.S.Dept of Commerce,*Measures for progress*, Washington, 1966, App B.

(2)   In the Archives d'Etat, Geneva, there are several references to this subject including Dossier 43 (ancien 214), containing notes on the revision of weights and measures from 1697 to 1748.

(3)   G.Bigourdan, *Le Système Métrique des poids et mesures*, Paris, 1901, pp 6,7.
H.Moreau, op cit, p 20, footnote, a mathematician, Simon Stevin, Bruges, in his treatise *De Thiende* (La Disme), Leyden, 1585, gave an explanation of decimal fractions and a decimal systems of weights, measures and money.

(4)   Address on 17.11.1913 by Alexander Siemens, *Bradford and the metric system*, pp 2,3, Verbatim report published by The Decimal Association, London.

(5)   NOTE
One of these official copies of the French metre found its way to a Swiss, Ferdinand Hasler, a mathematician and scientist. In 1805 on emigration to the USA, he took with him this 'Committee Meter' (as it subsequently became known in his new country) as well as copies of the kilogram and other French standards. In 1806, for economic reasons, he was obliged to sell them to the American Philosophical Society, of which he became a member in 1807. His copies of the French standards are now preserved in the archives of the United States National Bureau of Standards. Hasler, who had a distinguished career, became First Superintendant of the Coast Survey and of the U.S. Office of Weights and Measures. (*Measures for progress*, Rexmond C. Cochrane, U.S.Dept. of Commerce, Washington, 1966, p 529)

(6)   *Measures for progress*, op cit, pp 530,531

(7)   IEC Publication 164: *Recommendations in the field of quantities and units used in electricity*, Geneva, 1964, p 11

(8)   NOTE
At the end of December 1901, Prof Pietro Blaserna, the Italian member of the Bureau, wrote to Kelvin appealing for his help in obtaining an additional contribution of Fr 6484.- from Britain for the Bureau (agreed to by all other member bodies, and which was even smaller than the contribution of Italy) to enable work on the proposed revision of the kilogram and metre standards, etc.,to be undertaken. The British delegate, Mr Chaney of the Board of Trade, had abstained from voting on this matter. Prof Blaserna felt that although payment of this sum was a small amount for a country like Great Britain the loss for science could be considerable.
Kelvin immediately wrote asking Chaney the reason for non-payment, and suggesting that the Royal Society should try to induce the Government to authorize the payment. Chaney replied that the question of the British contribution 'towards this excellent Bureau' had not yet been considered and therefore Prof Blaserna's letter might be premature.
Kelvin apparently not satisfied, on January 13 this time wrote direct to the Prime Minister, Mr Arthur James Balfour, enclosing a copy of Blaserna's letter. Kelvin apologised for troubling him but 'that after some correspondence with the Board of Trade ....I feel that I am advised to do so. I have ascertained from Mr Chaney that he is throughly satisfied with the work of the International Committee, and that his reason for not joining in the small additional contribution asked for from England is that neither he (nor the Board of Trade, so I understand) can promise it without permission from the Treasury.'

Almost by return of post, Balfour replied on January 13 from Downing Street, that an official letter from the Treasury agreeing to the small additional contribution to the International Committee for Weights and Measures' was about to be sent.

Kelvin Collection, Cambridge University Library:

– DS 9,Letter [12].12.1901, Prof Blaserna(1836–1918) to Kelvin.
– DS 10,Letter, 23.12.1901,Kelvin to Blaserna.
– DS 11,Letter (copy),23.12.1901, Kelvin to Chaney
– DS 12,Letter (copy),26.12.1901, Chaney to Kelvin
– DS 1, Letter, 13.1.1902, Kelvin to Balfour
– DS 2, Letter, 18.1.1902, Balfour to Kelvin

Chapter 3
Metric controversy in Britain and the USA

The metric system for scientific purposes has now been adopted throughout the world. In Britain and North America, even today, however, certain sections of the community are loath to give up using the measuring system based on the inch, foot and pound system. Leading British newspapers such as *The Times*, and *The Daily Telegraph* are even, in 1991, printing readers' letters defending the old imperial system of weights and measures. (See Appendix C).[1]

In France even – where the metric system originated – its use throughout the country was not made obligatory until 1840. Similar legislation followed in other countries: 1856 in Switzerland, 1862 in Mexico, 1863 in Italy, 1866 legalised for scientific purposes in the USA, 1872 in Germany and Portugal, 1876 in Austria, and 1882 in Norway.[2, 3]

By the end of the 19th century, the metric system had been exclusively adopted by most civilised countries, with the notable exception of the United States and Britain. In 1864, Britain compromising with commerce and science authorised the use of the metric system, but concurrently with its imperial system of units. Two years later similar legislation followed in the United States. At the turn of the century, however, when the metric system became the subject of heated debate, Lord Kelvin's expertise in systems of measurement was called upon both in Britain and the United States where defenders of the old system of measurement were fighting a successful rearguard action against the 'revolutionary' metric system.

The British system, or lack of system, of measurement – which Kelvin called our 'insular and barbarous' system of weights and measures – was one of his pet aversions. Whenever the opportunity arose, he would deviate from his subject to criticize it.[4]

In 1865, he delivered a paper dealing with determinations of the values of Young's Modulus on long wires submitted to torsional vibrations. He adopted metric units for the calculations but could not resist adding a footnote:[5]

> 'It is a remarkable phenomenon, belonging rather to moral and social than to physical science that a people tending naturally to be regulated by common sense should voluntarily condemn themselves, as the British have so long done, to unnecessary hard labour in every action of common business or scientific work related to measurement, from which all the other nations of Europe have emancipated themselves.' [6]

Some twenty years later, while in Philadelphia he managed to work into a lecture on 'The wave theory of light' the following asides: 'You in this country are subjected to the British insularity in weights and measures...I look upon our English system as a wickedly brain-destroying piece of bondage under which we suffer'.[7]

The metric system in Britain was at various times the subject of government enquiry. In 1862, a Parliamentary Commission had recommended its introduction. Then followed several metric acts: the *Act* of 1864 intended to render the metric system permissible but which failed because of legal difficulties; then in 1878 an *Act* was passed making the use of the system legal for scientific purposes.

In 1895, an article in *Nature* discussed the changes considered necessary by the findings of a Parliamentary Select Committee in the British system of weights and measures. The need for a metrical system was emphasized to facilitate both foreign and home trade as well as the simplification that would result in the teaching in schools. The metric system had been applied with success in Germany, Norway, Sweden, Switzerland and Italy and many other European countries.

A Commission was engaged at this time in the United States in an investigation of a similar nature, and the Federal Government had earlier that year passed an *Act* rendering the metric system compulsory for pharmaceutical purposes.[8]

3.1 British Weights and Measures Acts

In Britain in 1897, a *Weights and Measures Act* was passed rendering permissible the use of the metric system of weights and measures.

This *Act* was referred to in 1904, when in the House of Lords debate on the second reading of the *Weights and Measures* [Metric System] *Bill*, Lord Kelvin (see also pp69–70) said:

'We have had nine years of permission to use the metric system without thereby rendering ourselves liable to punishment for a breach of the law, and experience has proved that the change from the system that has been so long in use in this country to a new system cannot be made over the whole country voluntarily. It is a case for compulsion, and I think the Legislation will be thanked by the country for having applied compulsion.'

At this point, drawing upon the contents of a letter (see Appendix A) he must have received only the previous day, from Professor Barr, Director of the James Watt Engineering Laboratory in the University of Glasgow, Kelvin continued:[9]

'While we are grateful to France for having given us the metric system, while we see France, Germany, Italy and Austria rejoicing in the use of it, it is somewhat interesting to know that, after all, the decimal system, worked out by the French philosophers, originated in England. In a letter dated 14th November, 1783,

18   THE DAILY TELEGRAPH, FRIDAY, NOVEMBER 9, 1990

The Daily Telegraph

181 MARSH WALL LONDON E14 9SR TEL : 071 538 5000 TELEX : 22874/5/6.
TRAFFORD PARK MANCHESTER M17 1SL TEL : 061 872 5939 TELEX : 668891

Measure for measure

 WO HUNDRED years ago, the French *Assemb-lée Nationale* set in train a process of revolutionary idealism for which mankind has paid dearly. Not content with turning the world upside down, France also laid hold of a quarter of its meridian and decreed that this should be the new way of measuring everything. One 10-millionth part of this length was declared to be the metre. Most of the other Utopian, rationalist, unhistorical — in short, insane — projects of these years came to naught. The renaming of the months, the redrawing of the calendar, the abolition of forms of address, all soon failed. But for some reason this particular folly endured.

Britain, as our Science Correspondent reports on another page, was invited to take part at the time, and sensibly refused. When this country finally succumbed, it did so for characteristically unideological reasons, but also, it must be sadly admitted, with an equally characteristic lack of concern for what people actually wanted. In 1964, the President of the Board of Trade was Mr Douglas Jay, who became notorious for declaring, in another context, that "the gentleman in Whitehall really does know best". Mr Jay gave a written answer in Parliament, saying that metrication in business would be allowed to proceed. In 1968, the Minister of Technology, Mr Tony Benn, a Jacobin if ever there was one, said it would be wonderful if *everything* went metric, although of course the change would be purely voluntary. Then, by degrees, came compulsion, first for medicines, then for schools, soon for much more, which was only stopped by Mrs Thatcher when she came into office in 1979.

Recently, the thing has begun again. The Robespierres of Brussels have ordered that we may not buy apples by the lb after 1994, that we may not travel in miles after the year 2000. When this shameful process is complete, it will be just to exclaim, as did Burke of the execution of Marie Antoinette: "The age of chivalry is dead, and the glory of Europe is extinguished forever." The British public is more likely to judge Mrs Thatcher's efforts in Brussels by her progress in saving our pubs from the litre than our banks from the ecu. Then the measure of her achievement will be imperial, not metric.

Paris tried avert metr mess in 179

By Roger Highfiel
Science Editor

THE muddle over metri imperial units could have avoided if the French Na Assembly had had its w 1790, when the foundatio the metric system were la

On the bicentenary o metre, Prof Harry Freem the University of Sa writes in the current is Nature on how the Asse started the process.

But for quirks of histo shows that the course of r cation might have smoother and that we have ended up with units on the swing of the pendu

In the wake of the F Revolution, the need rational system of mea ment was called for by al of the French parliam which gave the responsi to the Academy of Science

It decreed that the should write to the Kii England, in the hope the tem would be undertake the Academy and the Society in London.

"Presumably Louis who was not deposed 1792, did take some actio no record of any ensuing c spondence has turned says Prof Freeman.

The French decided in that a universal measure quarter of the Earth's m ian — should be the basis ten millionth of this l became the metre.

But a system based o length of a pendulum w swing of one second had suggested in the Anglo-F venture.

In 1798, France invited to participate but Britair then at war with France.

Editorial Comment — F

Fig. 2   Metric controversy

James Watt laid down a plan which was in all respects the system adopted by the French philosophers seven years later, which the French Government suggested to the King of England as a system that might be adopted by international agreement. James Watt's objects were to secure uniformity and so establish a mode of division which should be convenient as long as decimal arithmetic lasted.'

To illustrate how nonsensical was the British system of weights and measures, Kelvin then recounted how as a Captain in a Volunteer Regiment he had nearly been blown up as a result of having weighed out an incorrect charge of explosive during a laboratory experiment on a rifle. He had used the *apothecaries drachm* instead of the *avoirdupois drachm* as a measure to calculate the charge, but fortunately had discovered his error just in time to prevent the rifle exploding.[10]

Despite the Lords' favourable reception, the *Bill* on being returned to the Commons was referred to a Select Committee following its second reading.[11]

3.2 United States Congress and the Metric System

In a similar way, in the USA, during the first decade of this century, nine measures were presented to Congress intended to replace the measurement system by the metric or some other decimal system. Despite the testimonies of such experts as Kelvin and Alexander Graham Bell, none of the measures could be enacted.[12]

On April 24, 1902 Lord Kelvin gave evidence to the Committee on Coinage, Weights and Measures of the U.S. House of Representatives on the subject of the metric system. He made it clear that there could be no doubt that a change was essential in their measuring units.

In reply to a question regarding the possible universality of a system, he said: 'it seems perfectly obvious that it must be for the benefit of the world that the system of weights and measures should be world-wide'. Asked what would be the best system, he commented: 'that question the French philosophers and statesmen took under their very effective guardianship more than a hundred years ago, and with very great wisdom they chose a system that is almost ideally perfect'.

Later, he explained how he had had a running fight for a score of years with an old friend, Sir Frederick Bramwell, who was almost offended when Kelvin used the word 'inch' without attacking him. 'We cannot call the American or British measures of area and of bulk a system', he said, 'hardly even a jumble of systems . There is no system !'

At a certain point, Kelvin mentioned electrical units:

'The international system of electrical units, in which everything electrical is measured, is the same in America, Germany, France, England,etc., and all our instruments are founded on the centimetre and the gram....Every instrument-making workshop and engineering establishment on a large scale in England is now obliged to use two sets of standards in executing home and foreign orders....'

Towards the end of his evidence, Kelvin gave his view on the future prospects in England for the adoption of the metric system: 'I would rather that England should do it first and America should follow, yet I would very much prefer that America should lead if the end can be so accomplished sooner. And if America decides to make this reform England will follow very quickly'.[13]

Despite the recommendation of the Committee that the time had come 'for the retirement of our confusing, illogical, irrational system and the substitution of something better', the 1905 Bill intended to establish the metric system in all transactions amd activities of the Federal Government did not pass beyond the Committee stage.[14]

Both Britain and the United States continued to use their old measuring systems, except in scientific applications, despite two world wars, with all the confusion this caused with their metric-using allies. One particular cause for concern, was the lack of uniformity in screwthreads, a subject which had been investigated by Kelvin some fifty years previously but had never been completely resolved because of the metric question. (See Appendix B)

Notes

(1)   AUTHOR'S NOTE
This text has been prepared on an IBM P.C. incorporating an internationally recognised 3.5 inch disk.
(2)   *Measures for progress*, op cit, pp 527–530
(3)   NOTE
As late as 1964, the British Standards Institution issued a *Report* (PD5069) recommending 'the introduction of the metric system as the primary system of weights and measures in the United Kingdom within the shortest practicable period'.
(4)   *Life*, op cit, p 998
(5)   On the elasticity and viscosity of metals, *Roy.Soc.Proc*, XIV, 1865
(6)   *Life*, op cit, pp 435,436
(7)   ibid, p 808
(8)   *Nature*, 11.7.1895, pp 256,257
(9)   Letter, Prof Barr to Kelvin, 20.2.1904
Cambridge University Library (see Appendix B).
(10)   House of Lords, *Proc.* 23.2.1904
(11)   *The Morning Post*, 24.2.1904
(12)   *Measures for progress*, op cit, pp 209,210,534
(13)   Kelvin's full testimony appeared in supplementary hearings before the Committee of Coinage, Weights, and Measures, 24.8.1902 (L/C:QC91.U481). The above has been taken from a 7-page printed version of *Extracts from the evidence given by Lord Kelvin*, n.d.,n.p. (Cambridge University Library : Add 342 – PA250).
(14)   *Nature*, 12.6.1902, p 158

Chapter 4
Electrical measurements

Whereas in basic mathematics, the science of quantity, units are not of primary importance – the fundamental properties of arithmetic, algebra, geometry, trigonometry, dynamics, etc, are not dependent upon the existence of any particular set of units. For instance, the Pythagorean theorem relating the squares on the sides of a right-angled triangle, would be true in Euclidian geometry for any units of length.

The progress and extension of the electric telegraph made it essential for those responsible for the construction and operation of the lines to have a practical knowledge of the system. It became clear, however, that there was a large gap between the students' knowledge of the discoveries of Volta and Galvani, of Oersted and of Faraday, and the work of the practical electricians who had to obtain practical results by communicating their knowledge to others.

The development of electrical units and measuring standards from their inception in the 19th century is directly linked with advances made in scientific and engineering applications.

In those fields of science which are dependent upon accurate measurement and more particularly in applied science, the units of measure assume an important rôle. The confusion and misunderstandings resulting from a multiplicity of arbitrary units led to scientists of the stature of Lord Kelvin to seek a simple, and universal system of units.[1]

The fact that today the whole of the civilized world has adopted the *Système international d'unités (S.I.)* – likely to remain the primary internationally accepted system of measurement for a very long time – is directly linked to the pioneer work of Kelvin. The *S.I.* is a coherent system, that is to say, when two or more unit base quantities are multiplied or divided as required to obtain derived physical quantities, the result is unit value of the derived quantity.[2]

In a lecture on units of electrical measurement, Kelvin began with a statement which bears repetition:

> 'In physical science a first essential step in the direction of learning any subject, is to find principles of numerical reckoning, and methods for practicably measuring, some quality connected with it. I often say that when you can measure what you are speaking about, and express it in numbers, you know something about it; but when you cannot measure it , when you cannot express it in numbers, your knowledge is of a meagre and

unsatisfactory kind: it may be the beginning of knowledge but you have scarcely , in your thoughts , advanced to the stage of science, whatever the matter may be.' [3]

Kelvin regretted that to the non-scientific mind, accurate and minute measurement seemed a less dignified work than looking for something new. But, he emphasised, 'nearly all the grandest discoveries of science have been but the rewards of accurate measurement and patient long-continued labour in the minute sifting of numerical results':

- Newton's theory of gravitation was not the result of his sitting in a garden seeing an apple fall, but the result of a long train of mathematical calculations, which discovery he did not publish for years afterwards when he had learned of a serious correction of the previously accepted estimate of the Earth's radius.

- Faraday's discovery of specific inductive capacity, was the result of minute and accurate measurement of electric forces.[4]

- Joule's discovery of thermo-dynamic law through the regions of electro-chemistry, electro-magnetism, and elasticity of gases was based on a delicacy of thermometry which seemed simply impossible to some of the most distinguished chemists of the day.[5]

As Kelvin pointed out in this lecture, the earlier electricians had merely described phenomena such as attractions and repulsions, and flashes and sparks – the closest they came to measurement was the length of the spark. The first electrical measurements had been made in the 18th century by Robinson in Edinburgh, and of Coulomb in Paris, of electrostatic forces.

The idea of measuring electrostatic capacity in a scientific way was due to Henry Cavendish whose series of experiments measuring electrostatic quantities led to the general system of electrostatic measurement in absolute measure. With one or two exceptions, including the discovery by Faraday of the induction phenomenon, the whole theory of electrostatics was completed in the 18th century. Kelvin commented, 'It was merely left for us to work out the mathematical conclusions from the theory of Cavendish, Coulomb, and Robinson.' [6]

Oersted's great discovery of 1820 of the mutual force between a magnet and a current carrying wire, laid the foundation of electromagnetism.[7]

Ampère's discovery of the relations between two current-carrying wires rapidly followed, but by 1827 with his description of the laws of action of electric currents in his *Mémoire sur la théorie mathématique des phénomènes électrodynamiques, uniquement déduite de l'expèrience*, his creative contribution to the development of electrical science had virtually ended.[8]

In the same lecture, in speaking of the advances made in electrical measurement, Kelvin said:

'True, Cavendish and Coulomb last century, and Ampère, and Poisson, and Green, and Gauss, and Weber, and Ohm, and Lentz, and Faraday, and Joule, this century, had given us the mathematical and experimental foundation, for a complete system of numerical reckoning in electricity and magnetism, in electro-chemistry, and in electric thermodynamics; and as early as 1858 a practical beginning of definite electric measurement had been made, in the testing of copper resistances, insulation resistances, and electro-static inductive capacities, of submarine cables. But fifteen years passed after this beginning was made, and resistance coils and ohms, and standard condensers and micro-farads, had been for ten years familiar to the electricians of the submarine-cable factories and testing- stations, before anything that could be called electric measurement, had come to be regularly practiced in most of the scientific laboratories of the world.' [9]

Andrew Gray (Kelvin's assistant for many years) said that although Faraday was a non-mathematician, his own conception of the cutting of lines of magnetic force giving rise to the induced current, pointed the way to a quantitative investigation of current induction, but to render the mathematical theory explicit, and to investigate and test its consequences, required the highest genius and this is what Kelvin accomplished in the early years of his professorship at Glasgow University.

An enormous task confronted physicists at the middle of the 19th century by the discoveries of the preceding twenty-five years. Research work was hindered by the lack of precise knowledge of physical constants, which in turn was a consequence of the lack of exact definitions of quantities to be determined, but the biggest obstacle was the non-existence of any system of measurements. Units were arbitrary and depended upon apparatus in the possession of the individual experimenter which rendered them unsuitable for comparison with other results. [10]

Kelvin's biographer also emphasises Kelvin's contribution to the abstract science of electricity at this period. In 1846, Thomson, then aged 22, and already Professor of Natural Philosophy at Glasgow, appeared upon the scene. His mathematical approach to electrical matters enabled him to elucidate scientific principles as opposed to the more general expositions of Faraday with whom he corresponded on a number of occasions. [11]

The selection and realisation of electrical standards was an achievement of primordial importance to the world but one which has never received adequate acknowledgement. Kelvin's great accomplishment was to bring together all the experimental scientists of his time into one co-operative association for investigators whose individual efforts were aided by their combined results, expressed in a notation and described in language understood by everyone. [12]

In addition to his work on the establishment of accurate and reliable units of measurement, Kelvin insisted on the development of precision measuring instruments. As early as 1851 he placed orders for instruments with an optician, James White, and later became a partner in the firm which today is well-known in engineering circles as the Kelvin-Hughes Division of Smiths Industries.

In an article on Kelvin's measuring instruments, J.T.Lloyd has classified these in three separate categories: [13]

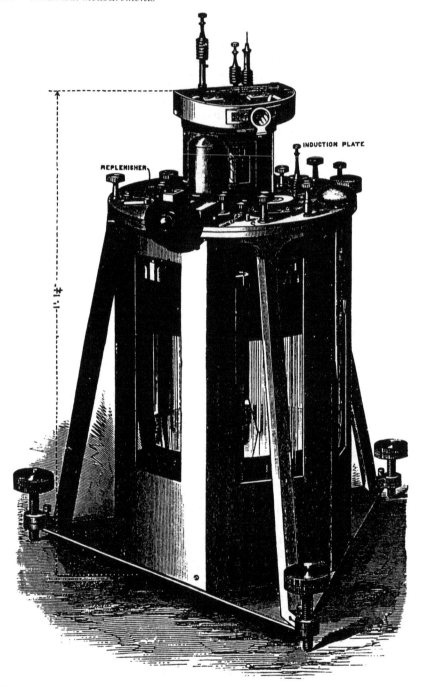

Fig. 3a Sir William Thompson's quadrant electrometer for measuring and comparing differences of potential

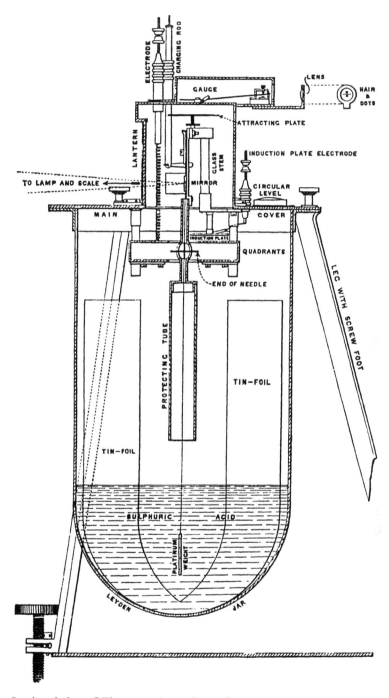

Fig. 3b Sectional view of Thompson's quadrant electrometer

- instruments which he invented, the attracted disk electrometer of 1855, to give just one example,
- those made for the needs of the telegraphic industry, in the 1850's,
- those required for test rooms and engine rooms from 1880 onwards.

Kelvin's great ingenuity, knowledge of practical engineering matters, including familiarity with mechanical devices, backed by his unrivalled experience in accurate electrical measurement enabled him to invent and direct the construction of measuring instruments.

His mirror galvanometer used for current measuring became extremely important in connection with the completion of the first Atlantic cable in 1858, since it was the only practicable method of receiving signals over long-distance cables. About ten years later, he invented the siphon recorder with which a permanent record of a received message was obtained on paper tape by ink from a fine siphon tube.

Kelvin's long series of patented electrical measuring instruments, which commenced in 1881, continued up to 1896. His graded galvanic potential and graded current galvanometers were effectively portable voltmeters and ammeters, respectively. These were followed up by meters and gauges for use on switchboards; his ampere gauges measured current from less than 1 A up to 6000 A.

He was able to put his knowledge of advanced electrical theory to practical applications, and at the same time to supervise the construction of prototypes in the workshop while attending numerous experimental trials in the laboratory. He did this without neglecting his many other scientific activities. (14)

Notes

(1)   Dr A.E.Kennelly, Historical outline of the electrical units, *Journal of Engineering Education*, Vol XIX, No 3, Nov, 1928, Harvard Engineering School, pp 229,230

(2)   B.Swindells, Understanding units of force, *Engineering*, 29.1.1971, pp 769–771

(3)   Sir William Thomson (Kelvin), Electrical units of measurement, Institution of Civil Engineers, 3.5.1883, *The practical applications of electricity*, London, 1884, p 149

(4)   NOTE
After completing the experimental part of a physical problem and requiring it to be treated by the mathematicians, Faraday would irreverently say "Hand it over to the calculators". Nevertheless, he insisted on the outmost accuracy in all measurements conducted during his experiments.(Address by William Spottiswoode FRS, *B.A. Report*, 1878, p 27)

(5)   Sir William Thomson, *B.A.Report*, 1871, pp xci,xcii.

(6)   Sir William Thomson, Electrical Measurement, *Popular Lectures and Addresses*, Vol 1, 1891, pp 430–439 (Address given on 17.5.1876)

(7) NOTE
Oersted's discovery in 1820 of the interaction between electricity and magnetism by the deflection of a magnetic needle by current in a wire had been published in Latin and sent to many scientists and learned societies in the world. In Geneva, Marc-Auguste Pictet reproduced the paper in French in his *Bibliothèque Universelle*, whose international value during the first part of the 19th century played an important role in the communication of scientific advances. Ampère visited Geneva and in Gaspard De la Rive's laboratory before 'Prevost, Pictet, de Saussure, Marcet, de Candolle, etc', verified Oersted's experiment. De la Rive recounts that Ampère, in his impatience to see the magnetic needle move, pushed it furtively with his finger; the needle then moved from left to right. De la Rive then readjusted one of the contacts and repeated the experiment but this time the needle moved from right to left. *Voila, cette fois*, said Ampère, quite innocently, *c'est bien dans ce sens que l'aiguille doit dévier: la première fois je l'avais poussée avec le doigt.*
J.L.Soret, *Auguste de la Rive*, Geneva, 1877, p.18, footnote.
A.Larseb, *La découverte de l'electromagnétism faite en 1820 par J.-C.Oersted*, Copenhagen,1920.p 27 (A footnote initialled 'A.'[Ampère], explains that the French translation of Oersted's Latin memoir had been communicated to him in Geneva by Pictet, and confirms De la Rive's report of the repetition of the experiment—except for the finger-pushing incident!).
(8) *DSB*
(9) Thomson, *Electrical units of measurement*, op cit, pp 82,83
(10) Andrew Gray,FRS, *Lord Kelvin:an Oration*, Glasgow, 1908, pp 16,17
(11) *Life*, op cit, pp 137–190
(12) Andrew Gray, op cit, pp 18,19
(13) J.T.Lloyd, Lord Kelvin and his measuring instruments, *Electronics and power*, 10.1.1974, pp 16–19.
NOTE
In 1899, in response to a request from Dr J.A.Fleming, a member of the B.A.Committee from 1883–1912, Kelvin sent him copies of a pamphlet 'Standard Measuring Instruments' (1899 Edition). Letter 5.3.1889, MS Add. 122/1/11C, Library University College, London.
(14) *Life*, op cit, pp 753–757
NOTE
As one example of his varied scientific interests, and considering only one organization, from 1860 onwards Kelvin served on 41 different committees of the British Association ranging for example from mathematical tables to screwthread design and elasticity of wires.
(*Life*, op cit, p 1127, footnote).

Chapter 5

Kelvin and British Association Electrical Units

By the middle of the 19th century, one of the biggest obstacles still confronting the development of natural sciences, and particularly physics, was the absence of a uniform system of defined measuring units. However, it was not until 1902 that Kelvin could assert that the international system of electrical units was the same in most modern countries and that all their instruments were founded on the centimetre and the gramme.[1]

In 1832, Carl Friedrich Gauss, a German mathematician, developed the concept of a comprehensive system of units which he referred to as 'absolute units'. Length, mass and time were the three fundamental units based on the millimetre, the milligramme and the second, respectively.[2]

An early British Association report in connection with the term 'absolute units', pointed out that *absolute* meant;

> 'that the measurement, instead of being a simple comparison with an arbitrary quantity of the same kind as that measured, is made by reference to certain fundamental units of another kind treated as postulates.'[3]

Professor Wheatstone in 1843, went a stage further in measuring work when he took for his unit one foot of copper wire weighing 100 grains, with the intention of adopting it as a standard for all resistances. It was Wheatstone also who first constructed measuring instruments which could be inserted in a circuit as multiples of the unit of resistance, which became known as rheostats or resistance coils.

This was soon followed by Poggendorff, Jacobi, Buff, etc., but each scientist employed a particular unit either of iron or of copper, or of silver wire. In the case of Pouillet, he had already – for his method was still officially in use at the turn of the century – related the resistance of various materials to that of mercury, adopting for the unit a column of mercury 1 metre high and 1 square millimetre in section.

In 1848, to enable scientists of various countries to compare their results, Jacobi sent each of them a length of wire, which became known as a Jacobi standard, to enable them to make reliable copies using using electrical means rather than by merely reproducing the length mechanically.[4]

The scientific world had to wait for nearly twenty years before another German, Wilhelm Eduard Weber, in 1851 extended this earlier work on units so as to embrace the entire electromagnetic and electrostatic systems of electrical units expressed in terms of the so-called absolute units.[5]

In his presidential address to the British Association in 1871, Sir William Thomson referred to the contribution of these two Germans:

'Great service has been done to science by the British Association in promoting accurate measurement in various subjects. The origin of exact science in terrestrial magnetism is traceable to Gauss' invention of methods of finding the magnetic intensity in absolute measure. I have spoken of the great work done by the British Association in carrying out the application of this invention in all parts of the world. Gauss' colleague in the German Magnetic Union, Weber, extended the practice of absolute measurement to electric currents, the resistance of an electric conductor, and the electromotive force of a galvanic element.

He showed the relation between electrostatic and electromagnetic units for absolute measurement, and made the beautiful discovery that resistance, in absolute electromagnetic measure, and the reciprocal of resistance, or, as we call it, "conducting power", in electrostatic measure, are each of them a velocity. He made an elaborate and difficult series of experiments to measure the velocity which is equal to the conducting power, in electrostatic measure, and at the same time to the resistance in electromagnetic measure, in one and the same conductor.

Maxwell in making the first advance along a road of which Faraday was the pioneer, discovered that this velocity is physically related to the velocity of light, and that, on a certain hypothesis regarding the elastic medium concerned, it may be exactly equal to the velocity of light. Weber's measurement verifies approximately this equality, and stands in science *monumentum oere perennius*, celebrated as having suggested this most grand theory, and as having afforded the first quantitative test of the recondite properties of matter on which the relations between electricity and light depend...' [6]

In December 1851, two of Thomson's papers were published in the *Philosophical Magazine*. The first, 'Applications of mechanical effect to the measurement of electromotive forces and of galvanic resistances in absolute units', advocated the extension and use of 'absolute units' as introduced by Professor Wilheim Weber that year for measurement of magnetism and electricity, which he restated in British measures, with the foot, grain and second as the fundamental units of length, mass and time.

He gave a table of measurements of resistances, showing considerable discrepancies, as made by different workers and advocated determinations of the absolute resistance of the same conductor by the direct method of Weber, and by the indirect method of calculating it by Joule's equivalent from the heat developed in the conductor by a current. His second paper 'On the mechanical theory of electrolysis' dealt with the absolute value of the electromotive forces concerned in electrolytic processes.[7]

In 1861, Sir Charles Bright and Mr Latimer Clark proposed the names of *galvat* for current, *ohma* for electromotive force, *farad* for quantity, and *volt* for resistance.[8]

This was taken up and Thomson – who after having advocated the general use of the absolute system for scientific investigation and for telegraph work over the previous ten years – succeeded, in 1861, in obtaining the setting up of a committee of the British Association on Electrical Standards.[9]

By this early recognition of the need for a sound scientific basis for units of measurement, and his lead in bringing about the subsequent international standardization of units, he laid the foundation for the phenomenal growth of the electrical engineering industry at the turn of the century.

When the committee was first appointed, there was no recognized coherent system of units for the measurement of electrical resistance, current, electromotive force, quantity or capacity. The primary task was to determine the most convenient unit of resistance and then to decide upon the best form and material for the standard representing that unit.[10]

After several meetings, the committee adopted as unit of resistance one based on the metre and second in the electromagnetic system of Weber, since one of the decimal sub-multiples of the metre-second system agreed within a few percent with the arbitrary mercury standards suggested by Siemens.

Thomson prepared some new resistance coils and sent them to Weber for checking. Furthermore, he devised a new method of determining resistance by means of a revolving coil but the experiments were not completed in time for inclusion in the first report of the committee.[11]

The summaries of the work of the British Association Committee, based on its published Reports from 1861 until 1869, were revised and reissued as one volume in 1873 by Sir William Thomson, Dr J.P.Joule, Prof J.Clerk Maxwell and Prof Fleeming Jenkin, who acted as editor.[12]

A preface to this volume includes the following statement:

> 'As the centimetre is now frequently used instead of the metre as the fundamental unit of length, it has been introduced into the Reports, not to the exclusion of the metre, but along with the metre, all results being expressed in terms of both measures.'

By the time of the B.A. Committee's report in 1869 the absolute system had been launched for general use.

5.1 First British Association Report: 1862

The first *B.A. Report* of October 3, 1862, summarised the two distinct questions before the committee. They had to determine first, the most convenient unit of resistance; and second, the best form and material for the standard representing that unit. While they had reached a provisional conclusion on the first question they had been unable to arrive at any conclusion for the second question.

In their determination of a unit they set themselves a number of objectives:

- The magnitude of the unit should be suitable for the more usual electrical measurements, without using large numbers of cyphers or series of decimals.
- It should be related to units which may be adopted for the measurement of electrical quantity, currents, and electromotive force, i.e. it should form part of a complete system for electrical measurements.
- The unit of resistance should like the other units of the system bear a definite relation to the unit of work.
- The unit should be reproducible with exactitude, in order that if the original standard were destroyed it might be replaced, and also so that observers could manufacture copies of the standard.
- The unit should be based on the French metrical system rather than on that now used in Britain.

In view of the second requirement, having a coherent system of units, the Committee was obliged to determine not only the unit of resistance but units for current, quantity and electromotive force.

This first *Report* also dealt with the value of the unit of resistance and incidentally established the absolute system of electrical units. The relationship between the various units, was summed up as follows:

> 'A battery or rheomotor of unit electromotive force will generate a current of unit strength in a circuit of unit resistance, and in the unit of time will convey a unit quantity of electricity through this circuit and do a unit of work or its equivalent.' [13]

5.2 International cooperation 1862

On April 3, 1862 the B.A.Committee sent out a circular 'to Foreign Men of Science' in leading universities or institutions. The circular signed by Fleeming Jenkin on behalf of the six-member committee, comprising also Thomson and four other Fellows of the Royal Society (Professors Williamson, Wheatstone, Miller, and Matthiessen), announced the setting up of the British Association Committee on Electrical standards of resistance, and called for suggestions to further the scope of its work.

The foreign recipients of the circular were Prof Edlund (Upsala), Prof Fechner (Leipsig), Dr Henry (Washington), Prof Jacobi (St Petersburg), Prof Kirkhoff (Heidleberg), Prof Matteucci (Turin), Prof Neumann (Königsburgh), Prof Poggendorff (Berlin), M.Poulain (Paris), Dr Werner Siemens (Berlin), Prof Weber (Göttingen). The proposed forms of standard units were followed by a detailed discussion of the arguments for and against each:

i) A given length and weight or section of wire made of some pure metal, and observed at a given temperature, as originally proposed by Profs Wheatstone, Jacobi and others.
ii) Units based on Weber's and Gauss' system of absolute measurement.
iii) A given length and section of pure mercury at a given temperature.

This important communication pointed out that the timing of the work was considered favorable since 'no local units have as yet taken root'. Fleeming Jenkin requested details of their own work in the field, with references, to assist him in preparing an historical summary of the various units proposed.

Two replies were reproduced from Prof Kirchholl and Dr Siemens: the former considered the unit proposed by Weber was most suitable, and that the column of mercury should be established as the unit of resistance. Dr Siemens also felt that the column of mercury should be adopted. Both replied giving historical references for their views.

5.3 Second British Association Report: 1863

The second *B.A. Report* issued in 1863, stated that the original terms of reference had now been extended to cover not only the best unit of resistance but 'the much larger subject of general electrical measurement'. On the question of electrical measurements, the report mentioned Prof Thomson's standard gauge for measuring electromotive force or difference of potential and also his electrometer both of which would become as necessary to all practical electricians as standards of resistance and sets of resistance-coils.

This report also included the famous paper by Prof Maxwell and Fleeming Jenkin 'On the elementary relations between electrical measurements'.

5.4 Third British Association Report: 1864

The third *B.A. Report* in 1864, after confirming 'the adoption by the committee of the absolute electromagnetic system of measurement, based on the metre, gramme and second, with certain modifications to facilitate the practical construction or use of the standards', said that no standards based on the 1863 determination had been 'officially issued, since a second determination was considered essential before complete dependence could be placed either on the method employed or in the results obtained'.

Although no reply had been received from France, assurances had been received from Britain, India, Australia and Germany that the British Association system of units would be readily adopted and that it would also be accepted in America and in many other parts of the world.

This *Report* later resulted in the issue of the standard of resistance called the *B.A.Unit*, or *ohmad*, afterwards shortened to *ohm* (see committee's fourth report below).

5.5 Fourth British Association Report: 1865

It was not, however, until their fourth meeting held at Birmingham in September 1865 that the Committee, now expanded to 12 members,

Fig. 4   British Association standard ohm, purchased by Faraday in 1865

reported that 'the unit of electrical resistance had been chosen and determined by fresh experiments and that the object for which they had first been appointed had now been accomplished'.

The standard made of platinum-silver alloy coils was distributed to thirteen overseas recipients, or sold – the report added, 'Dr Faraday, on behalf of the Royal Institution, was the first purchaser' (Faraday's standard is shown in Figure 4). It also stated that 'the new unit has been actually employed to express the tests of the Atlantic Telegraph Cable'.

A report by Thomson on electrometers and electrostatic measurements was appended to the *B.A.Report*. In addition, a paper by J.P.Joule FRS dealt with the determination of the dynamical equivalent of heat from the thermal effect of electric currents. Joule's report begins:

> 'Sir W.Thomson, as long ago as 1851, showed that it was desirable to make experiments such as are the subject of the present paper. They have necessarily been delayed until a sufficiently accurate method of measuring resistance was discovered. Such a method having been described by Sir William, and carried out into practice by Professor C.Maxwell and his able coadjutors, the task assigned to me by the Committee of Electric Standards was comparatively simple.'

At that time, apart from electroplating, the only industrial application of electricity was telegraphy. Nevertheless, the theory of electricity was well developed and covered all the fundamental laws including Ohm's law, the laws of electromagnetic induction as well as those of electrostatic induction. There were galvanometers, batteries and resistance coils, but their calibration was crude and the practical application of the laws was far from adequate.

As regards electrical units, a comparatively recently found letter from one of the B.A. committee members, C.F.Varley, written in 1865 to Thomson shows how complex the unit question remained. (Clarke, as mentioned above, who had already in 1861 discussed new names for units with Sir Charles Bright, seems to have devoted considerable energy to thinking up new names for units !).

Varley's new proposal was as follows:

'Latimer Clarke and I the same evening, and Fleeming Jenkin and I the next evening, discussed the matter. Latimer Clarke proposed the following:

	1 unit	1 million units
Potential	*Galvad*	*Galvon*
Resistance	*Ohmad*	*Ohmon*
Current	*Voltad*	*Volton*
Quantity	*Farad*	*Faron*

...Jenkin objected to both, and one objection is a good one (and will apply to me and to you) *viz:* he writes so badly that if the magnitude be expressed by a termination, *Ohmad* and *Ohmon* will be confounded in indiscreet writing....I should like to introduce a French name into the list. We have Germany, England, and Italy represented: suppose *Ampère* were used to represent the magnetic pole, *viz* the unit pole?....I should like to have seen Weber's or your name introduced, attached to the unit of potential and have suggested it: fear has been expressed lest we should use any name on the committee and so give opponents an opportunity of saying the "unpleasant". I object to *Galvad* because Galvani discovered next to nothing, ...' [14]

5.6 Sixth British Association Report 1869

Appended to the brief 1869 *Report* were two papers, the first by W.F.King: 'Description of Sir Wm. Thomson's experiments made for the determination of v, the number of electrostatic units in the electromagnetic unit' based on his absolute electrometer and electrodynamometer, which had been referred to at the previous meeting, and the other by Maxwell dealing with the same subject, 'Experiments on the value of v, the ratio of the electromagnetic to the electrostatic unit of electricity'.

This report 'fairly launched the absolute system for general use; with arrangements for the supply of standards for resistance coils in terms of a unit, first called the British Association unit, and afterwards the *Ohm;* of

Fig. 5   Kelvin in 1852 (From S.P. Thompson's *Life of Kelvin*)

which the resistance was to be, as nearly as possible, 10,000 kilometres per second.'[15]

5.7 Subsequent work of British Association on electrical standards

Thomson pointed out in a lecture given in 1876, that a revision of the B.A. unit was now being undertaken in terms of the absolute scale of centimetres per second. According to Joule, on the one hand, the B.A. unit, the Ohm, was too small; on the other hand, in Germany, Kohlrausch considered that the Ohm was a little on the other side of the exact thousand million centimetres per second. Kelvin held that he would not be satisfied until both points of view were satisfied.[16]

Hand in hand with the establishment of accurate and reliable units of measurement, he insisted on the development of precision measuring instruments. Thomson's biographer succinctly summed up his contribution throughout all this long work: 'Thomson was the inspiring force in the committee, incessant in interest, and fertile in suggestion.'[17]

The absolute system remained in general use in Britain and the USA for a number of years before the practical adoption of the absolute system by France, Germany and other European countries at the International Electrical Congress held at Paris in 1881, the subject of the next chapter.

Immediately before the Congress, in July 1881, a director of the Swiss Telegraphs' Department pointed out that the absolute units of electricity were too complicated and abstract to define and that manufacture of the relevant standards was extremely difficult. As a result, the most celebrated experimenters could not agree on the results of their work. He felt that the values of the *weber, volt, ohm* and *ampere* were still unclear and impossible to reproduce accurately.[18]

This state of affairs also led Hospitalier, the somewhat temperamental editor of *L'Electricien* in Paris, to complain bitterly just before the Congress that the ten different units for current, twelve for voltage and fifteen for resistance constituted *une véritable tour de Babel* which electrical engineers regarded with horror and disgust.[19]

The work of the British Association Committee on Electrical Standards continued – with Lord Kelvin remaining an active member until his death in 1907. Meetings continued to be held annually until 1912. During the latter years of its existence, the Committee was active in promoting international uniformity in standards and many experiments were made at the National Physical Laboratory on behalf of the Committee.

In the history of this sector of British Association activities (published in 1913) it is stated that the appointment by the London Conference of 1908 of an International Scientific Committee of fifteen scientists to direct work in connection with the maintenance of standards relieved the Committee of much of its responsibility.

In the words of the editors of the above history, the main objects for which the B.A.Committee had been appointed had thus been achieved; in all the principal countries of the world the same units of resistance, of current, and of electromotive force had been adopted and the standards in use were practically identical.[20]

Notes

(1)   Lord Balfour *Kelvin Centenary Oration and Addresses*,London, 1924,p 27

(2)   The term 'absolute' introduced in his paper *Intensitas vis magneticae terristris in mensuram absolutam revocata* in 1832 was simply to indicate that this system of units was independent of instruments or gravity.

(3)   *B.A.Report*, 1863,p 112 (Committee members included Profs Thomson, Wheatstone, Siemens, Maxwell, Joule).

(4)   E.E.Blavier, *Des grandeurs électriques et de leurs mésure en unités absolues;* Paris, 1881, pp 3–7

(5)   Michael Wooley, *Telecommunications Journal*, Vol 48, ix, 1981, pp 543–547

(6)   Sir William Thomson FRS, *B.A.Report* 1871, pp xcii,xciii
      NOTE
      A further reference to the work of Gauss is given in Sir William Thomson, Electrical Units of Measurement,*The Practical Applications of Electricity*, I.C.E,London , 1884, pp 156,157.

(7)   *Life*, op cit, p 227

(8)   *Electrician*, vol i, p 3,9.11.1861 (*Life*, pp 417,418)

(9)   Thomson, *Electrical units of measurement*, op cit, p 153
      NOTE
      His paper 'On the electric conductivity of commercial copper' *R.S.Proc.*,VIII,1857, employed the absolute units giving the results of his experiments and comparisons with the standard wires used by Weber,Kirchoff and Jacobi. (*Life*, p 341).

(10) R.T.Glazebrook and F.E.Smith,*Introduction to the collected Reports of the Committee on Electrical Standards*, Cambridge, 1913.

(11) *Life*, op cit, p 418

(12) Fleeming Jenkin, *Reports of the Committee on Electrical Standards appointed by the British Association for the Advancement of Science*, (A record of the history of absolute units and of Lord Kelvin's work in connexion with these), London, 1873.

(13) R.T.Glazebrook, *Proc.I.E.E.*, 1907, p 7. Glazebrook commented: 'the words sound very simple and elementary in 1906'.

(14) Letter from C.F.Varley in Kelvin collection Univ. of Glasgow. J.T.Lloyd, Lord Kelvin and his measuring instruments, *Electronics and Power*, 10.1.1974, p 18
      NOTE
      Mr Lloyd at the time of writing this article was Keeper of the Kelvin Collection at Glasgow University. Varley's letter was reprinted with the permission of the Court of the University of Glasgow.

(15) Sir William Thomson, *The practical applications of electricity*, London, 1884, pp 153,154

(16) Kelvin, Electrical measurement, *Popular Lectures and Addresses*, Vol 1, 1891, pp 430–455 (Address given 17.5.1876).

(17) *Life*, op cit, pp 417–420,1127 footnote

(18) The Journal one hundred years ago, *Telecommunications Journal*, Vol 48, VII/1981, p 442.

(19) *L'Electricien*,1.7.1881, p 313

(20) Fleeming Jenkin, op cit, pp xxii,xxiii

Chapter 6
International Electrical Congresses

6.1 First International Electrical Congress: Paris, 1881

At the meeting of the British Association held in York in 1881, with Sir William Thomson present, the delegates discussed the forthcoming International Electrical Congress to be held in Paris. The firm hope was expressed 'that other nations might agree to the system of units and standards which the British Association Committees had evolved, chiefly under Thomson's inspiration since 1862'.[1]

That autumn, Thomson crossed over to Paris as one of the British members forming part of the 250 delegates from 28 countries represented at the Congress, which lasted from 15 September to 19 October 1881.

The Electrical Exhibition held at the same time as the Congress was an outstanding success and drew electrical contractors from all parts of the world including the United States. For the first time in Europe the nearly one million visitors were able to see all that the electrical industry could provide in the way of lighting, telegraphy and telephone systems.

The following distinguished scientists were present:

Germany	:	von Helmholtz, Clausius, Kirchhoff, Werner Siemens.
Austria	:	Ernst Mach
Belgium	:	Eric Gerard, Zénobe Gramme, Gilbert
U.S.A.	:	Rowland
France	:	J-B Dumas, Mascart, Planté, Marcel Deprez
Britain	:	Lord Raleigh, Sir Wm.Thomson, Wm.Crookes, Hopkinson
Italy	:	Govi, G.Ferraris
Norway	:	Broch
Russia	:	Lenz
Sweden	:	Thalen
Switzerland	:	H-F.Weber

Sir William Thomson, as one of the three foreign Vice-Presidents appointed, was a notable figure throughout the Congress. The Secretary of the section dealing with electrical units, which met on 16 and 17 September, was Prof E.Mascart (see also page 3 and its associated Note 11), who remained a close friend and colleague of Thomson throughout his career particularly in their work connected with the International Electrical

Congresses and electrical standards. Mascart also attended all the succeeding International Congresses including the all-important Congress of St.Louis in 1904 and that of 1905.

Prof Mascart recounts how the delegates enjoyed the unforgettable scenario of Thomson and Prof. Helmholtz (Germany) disputing hotly in French, each with his own distinctive style of pronunciation. Mascart was Secretary of the section dealing with electrical units which met on September 16 and 17, 1881, and in an off-the-record account he recounted how agreement was finally reached on international units.

Throughout the first day, the delegates reviewed the questions before them. On the second day, they concentrated on whether the units would be based on the British Association unit of resistance or the arbitrary Siemens unit, a column of mercury one metre long. The discussion was involved; proposals and objections came from delegates quite unaware of the real nature of the resolutions to be adopted. Finally at about 4.30 p.m in the afternoon, the chairman, M.Dumas, decided to adjourn the meeting. That evening, Dumas confided to Mascart that the reason for his suspension of the meeting was because he felt certain that no decision would be reached.

The next morning Mascart found a card from Thomson waiting for him – 'Hôtel Chatham 6.30 p.m.' The subsequent private meeting in an ante-room of the hotel was an imposing one (recounted Mascart): there were Lord Kelvin and William Siemens representing England; then Helmholtz, Clausius, Kirchhoff, Wiedemann and Werner Siemens representing Germany.

After considerable hesitation, Werner Siemens eventually accepted the proposed solution provided that the system of measurements would be qualified 'for practical use'. Mascart found no objection to this and in pencil wrote out the draft of the convention on the top of the piano.

The practical system had for base the C.G.S. electromagnetic units. The *ohm* and the *volt* were defined, it being left to an international committee to lay down the dimensions of the column of mercury to represent the [B.A.] ohm.

Late that evening, Mascart visited the hotel room of M.Dumas (who was Permanent Secretary of the Academy of Sciences) and told him of the agreement on the electrical units. Mascart felt that the credit was due partly to the firmness of Dumas, and partly to the influence on Werner of his brother William Siemens, who had been engaged on the initiative of the British Association.

As they still had only two units, the *ohm* and the *volt*, and it was necessary to complete the system, Mascart asked M.Cochery, (the French Minister of Posts and Telegraphs) who presided over the Congress, whether the committees could continue their work.

As the reply was negative, Mascart accompanied by Sir William and Lady Thomson, took a hot chocolate in the Restaurant Chibest near the Congress Hall. Seated at their marble-topped table, it was this small committee which agreed on the units *Ampere* (instead of *Weber*), *Coulomb*, and *Farad*.

When Mascart duly read out the text on September 21 to the committee

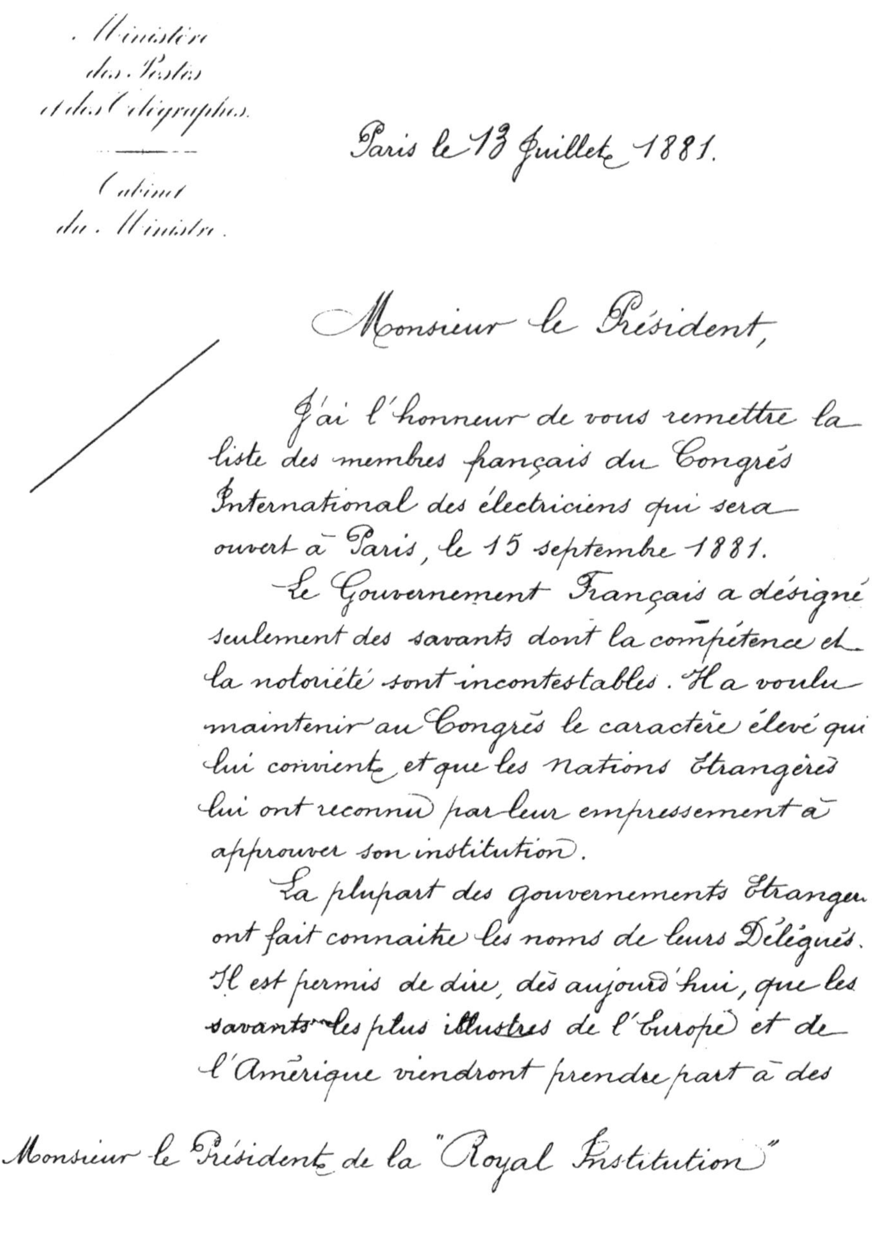

Ministère
des Postes
et des Télégraphes.

———

Cabinet
du Ministre.

Paris le 13 Juillet 1881.

Monsieur le Président,

J'ai l'honneur de vous remettre la liste des membres français du Congrès International des électriciens qui sera ouvert à Paris, le 15 septembre 1881.

Le Gouvernement Français a désigné seulement des savants dont la compétence et la notoriété sont incontestables. Il a voulu maintenir au Congrès le caractère élevé qui lui convient, et que les nations étrangères lui ont reconnu par leur empressement à approuver son institution.

La plupart des gouvernements étrangers ont fait connaître les noms de leurs Délégués. Il est permis de dire, dès aujourd'hui, que les savants les plus illustres de l'Europe et de l'Amérique viendront prendre part à des

Monsieur le Président de la "Royal Institution"

Fig. 6 Invitation from M. Cochery, President of the Congress, to the Royal Institution

discussions qui contribueront puissamment
à l'avancement des sciences physiques.

Je n'ai fait préparer aucun programme
de la session du Congrès. J'ai pensé qu'il était
équitable et libéral de laisser celui-ci déterminer
lui-même l'ordre et l'espèce de ses travaux.

Le Gouvernement de Sa Majesté
Britannique a nommé comme Délégué
de la Grande Bretagne Monsieur le Comte
de Crawford et Balcarres, Pair d'Écosse et
du Royaume-Uni. D'autre part, M. M. les
ingénieurs Graves et Preece ont été désignés
par l'Honorable Post Master Général pour
représenter son Administration.

J'ai l'honneur de solliciter que la
' Royal Institution ' veuille bien aussi
désigner ses Délégués.

Agréez Monsieur le Président, l'assurance
de ma très haute considération.

Le Ministre des Postes et des Télégraphes
Président du Congrès International des Électriciens

Fig. 7   Electrical dynamos at the 1881 Paris Exhibition

members – a number of whom were quite unaware of the *ad hoc* meeting on the Saturday morning – they were very surprised but as the observations of Thomson and Helmholtz left no room for any hesitation, the practical system of units was thus born.[2]

The following seven resolutions were adopted by the Congress on September 22, 1881: the official French text is reproduced first:

1. On adoptera pour les mesures électriques les unités fondamentales: centimètre, gramme-masse, seconde (C.G.S.).
2. Les unités pratiques, l'*ohm* et le *volt*, conserveront leurs définitions actuelles: 10^9 pour l'ohm et 10^8 pour le volt.
3. L'unité de résistance (*ohm*) sera représentée par une colonne de mercure d'un millimètre carré de section à la température de 0°C.
4. Une Commission internationale sera chargée de déterminer par de nouvelles expériences, pour la pratique, la longueur de la colonne de mercure d'un millimètre carré de section à la température de 0°C qui représentera la valeur de l'ohm.
5. On appelle *ampère** le courant produit par un volt dans un ohm.
6. On appelle *coulomb* la quantité d'électricité définie par la condition qu'un ampère(*) donne un coulomb par seconde.
7. On appelle *farad* la capacité définie par la condition qu'un coulomb dans un farad donne un volt.[3]

English version (several versions exist):

1. the basic units used for electrical purposes shall be the centimetre, the gramme and the second (CGS);
2. the present definitions of the practical units *ohm* and *volt* shall be retained at 10^9 for the ohm and 10^8 for the volt;
3. the unit of resistance (*ohm*) shall be represented by a column of mercury with a cross-section of one square millimetre at a temperature of zero degrees centigrade;
4. an international committee shall be requested to conduct new tests to determine for practical purposes the length of the column of mercury with a cross-section of one square millimetre at a temperature of zero degrees centigrade which will represent the value of the ohm;
5. the current produced by one volt in one ohm shall be called an *ampere;*
6. the quantity of electricity defined by the condition that one ampere produces one coulomb per second shall be called a *coulomb;*
7. the capacity defined by the condition that one coulomb in a farad produces one volt shall be called a *farad.*

In his comments on the resolutions, Sir William Thomson said:

'...The committee wishing to incorporate in the system the names of Ampère, the founder of electrodynamics, and Coulomb, to whom we owe the early definitions and the establishment of the science of electrostatics, now proposes to give the names of *ampere* and *coulomb* to measures of current and quantity of electricity, irrespective of the time defined above in the CGS system. Finally the name of Faraday will also be maintained in the *farad* as a measure of capacity...' [4]

6.2 International Electrical Conference for the Determination of Electrical Units: Paris, 1882–1884

The International Conference for the determination of electrical units convoked to meet at Paris in October 1882, decided that on the basis of the determinations made there was insufficient agreement to specify a value for the *ohm* reckoned as a column of mercury. The French Government was asked to make available the necessary standards of resistance to the Governments represented to enable scientists to undertake further research so as to facilitate comparisons.

In a lecture to the Institution of Civil Engineers the following year, Thomson referred to the definitive practical adoption of the absolute system as decreed by the International Conference in October 1882:

The decision adopted was, not to take the British Association unit. Doubt had been thrown upon its accuracy, which we shall see was well founded. The question of a strict foundation for a metrical system was before the Conference, and it was inclined to adopt the absolute system, but the question occurred "What is the *ohm* ?" Who can see an *ohm* ? Who can measure the resistance of any conductor for us, in this absolute measure of Weber's ? Weber's own measurement greatly differed from that of the British Association....The answer adopted by the Conference was to ask for a definition of an absolute system in terms of a column of mercury. The column of mercury was the one standard in existence, that could be reproduced otherwise than by merely copying from one wire to another; and it was naturally adopted as the foundation upon which a standard, if not a practical unit to be used, should be founded.[5]

The International Conference which met again in April/May 1884 decided that the legal *ohm* was to be the resistance of a column of mercury 106 cm in length and of section 1mm^2 at a temperature of melting ice. This recommendation was to be transmitted to the various Governments for international adoption.

At the same time, it recommended the construction of primary mercury standards and also the use of secondary solid resistances of metal alloy which should be frequently compared together and also with the primary mercury standard. The *ampere* to be the current whose absolute value was 10^{-1} C.G.S. electromagnetic units. The *volt* to be the electromotive force to maintain a current of one *ampere* in a conductor whose resistance was one legal *ohm*. The legal system thus adopted was not to prejudice future decisions since it would remain in force for 10 years only.[6]

6.3 International Electrical Congress: Paris, 1889

For Sir William Thomson, the year 1889 must have been a full one. In addition to his professorial and other activities he had been elected President of the Institution of Electrical Engineers for that year. In the spring, he had lectured at the Royal Institution, London, on electrostatic measurement. On July 1 he read an important paper to the Royal Society of Edinburgh on the molecular constitution of matter.

Early in June, M.Mascart wrote asking Thomson to attend the Congress. This invitation was politely refused on the grounds of Thomson's duties as President of the Institution, and also because of the establishment of an electrical standardizing laboratory which, they hoped, would be 'taken in hand by our Government'.

Mascart insisted once again, however, at the end of the month, and this time Thomson gave in. In his letter of acceptance of June 23, he wrote:

> 'I cannot resist your letter of the 18th, kindly insisting on my presence at the Congress of Electricians, and I have therefore arranged to come to Paris in time to attend at the opening meeting on the 24th of August, and to remain for, at all events, several days. Lady Thomson will be with me, and looks forward with pleasure to seeing you and Mme Mascart again in Paris, since you will not come to see us here this year.[7]

The International Electrical Congress which opened in Paris on August 24, 1889 cannot be considered as official since there were no recorded French or Foreign Government representatives among the 600 members present. Prof Mascart was President, and Sir William Thomson, Honorary President, who, while in Paris, presented three short scientific papers to the Académie des sciences.

Agreement at the Congress was reached on the following definitions of units:

- the practical unit of work is the *joule*. It is equal to 10^7 CGS units of work. It is the energy expended during one second by one ampere in a resistance of one ohm.
- the practical unit of output is the *watt*. It is equal to 10^7 CGS units of output. The *watt* is equal to one joule per second.

It was also decided that for industrial applications, the practical unit for coefficient of induction is the *quadrant*. One *quadrant* $= 10^9$ centimetres.[8]

According to the American scientist, Dr A.E.Kennelly, the names 'maxwell' and 'weber' were recommended in the practical system for the units of magnetic flux and magnetic flux density, respectively, but no action was taken.[9] The Congress defined the *joule* and the *watt*, however, not only in terms of the CGS system but also in terms of the practical system. It was unanimously decided that for industrial practice, the output of machines would be expressed in kilowatts instead of in horsepower.

In his closing speech, Prof.Mascart made an important review of the role of electricity in the world. On behalf of all those present, and to their applause, he expressed their appreciation to Sir William Thomson, who was able to combine the highest intellectual speculations with the study of practical applications. He had transformed all their electrical measuring instruments, sustained the courage of those involved in the gigantic transatlantic cable enterprise while at the same time conceiving transmitting and receiving apparatus which after 30 years of operation had not required any modifications. His compass and sounding apparatus now adopted by all the navies of the world, constituted a great service to mankind. Sir William Thomson, he said, whose name was on everyone's lips, and in the hearts of all those who had had the privilege of associating with him, had guided the British Association in proposing the reforms, since confirmed by the Congress.

Prof Mascart then spoke of the extent that electricity was affecting social life and human resources and penetrating all the experimental sciences, and whose precise terminology had resulted in great progress in their understanding of the phenomena in nature.

The different units and their values constantly employed in practice contributed to the spread of scientific knowledge, enabling concepts to be expressed in words, which was a direct application of the indivisible link between language and thought so vital to man's intellectual development.[10]

In a reference to the 1881 International Congress, he emphasised the altogether unexpected and tremendous advances in electricity, and how it had taken 40 years before the means had been found to use alternating current – first conceived at the hands of Faraday direct from his discovery of induction – in direct applications, thanks to Messrs Pacinotti and Gramme.

Shortly after the Congress, Sir William Thomson was made a Grand Officer of the *Légion d'honneur*.[11]

A report in *L'Electricien* the following year, stated that the name *quadrant* had been agreed at the Congress for the coefficient of induction, without opposition from the British and the Americans. But the British were continuing to call it the *secohm*, while the Americans desirous of having a name from one of their scientists insisted on referring to it as a *henry*.[12]

Subsequent to the Paris 1889 Congress, unofficial independent international congresses or conferences were held: at Frankfort in 1891, Edinburgh 1892, and Geneva in 1896.[13]

6.4 International Electrical Congress: Chicago, 1893

In the USA, by about 1885, the telephone and electrical light had become commercial realities; the first commercial electric trolleyway operated that year and the first electric power-plant not long afterwards. This technological progress was hampered, however, by the lack of precise units and measurements. By the time of the Chicago Congress in 1893 the need for some sort of agreement on electrical measurements had become imperative and various values were adopted for the basic units. The United States subsequent to the Congress became in fact the first Government to enact the definitions and values of the adopted units into law on July 12, 1894.[14]

LORD KELVIN.

Fig. 8 Pencil illustration of Lord Kelvin

According to Dr A.E.Kennelly, the Chicago Congress of 1893 was notable in the history of electrical units in that it formulated specifications for the standard *ohm* to a precision of .1 % as a result of the steady accumulation of measurements of pre-existing resistance standards in absolute measure. Its decisions formed the basis of legislation on electrical units and standards in all parts of the world.

Prior to the Chicago Congress, changes in the international electrical units might have been possible because not many countries had enacted laws on their electrical standards. After the Chicago Congress, however, changes could only be introduced with difficulty since the units and standards had been entered on the statute books of many countries. This was because the Congress had addressed its resolutions to the various governments represented at Chicago recommending them to 'formally adopt as legal units of electrical measure' the following units: the *ohm, ampere, volt, coulomb, farad, joule, watt,* and *henry* (substituted for the term *quadrant*), all in the practical system.

The *ohm, ampere, joule* and *watt* were defined on the basis of the corresponding units of the CGS system, with standards according to specifications for the *ohm* and *ampere,* but with reference to the practical system, for the *joule* and *watt.* In the case of the *volt,* the unit was defined with reference to the practical system while its standard was defined with reference to a specified voltaic cell. The *coulomb, farad* and *henry* were defined in terms of the practical system only.

These units (except for the *henry,* which as mentioned above, was a new name for *quadrant* of the preceding Paris Congress) were replications of the units and unit names already adopted internationally, but differed slightly in magnitude owing to the changed specification for the new standard ohm. The Chicago-Congress series of units was therefore called for distinctiveness, 'The International Series of Units', with individual units referred to as the *international ohm, international volt,* etc.[15]

The Chamber of Delegates adopted the *international ohm* of 106.3 cm of mercury, as an amendment of the legal ohm of 106 cm in 1884, which was an amendment of the B.A. ohm of 104.8 cm in 1875.

The *international volt* was defined in terms of the ohm and ampere fundamentally, but also collaterally as a fraction of the e.m.f. of a standard Clark cell, the specifications for the preparation of which was left to a committee which, according to Kennelly in 1904, had yet to submit a report.

The *international ampere* was defined in terms of the rate of deposition in a standard silver voltmeter. A table of international notation was accepted and printed as an appendix to the report, which was nearly 500 pages long.[16]

Louis Joly emphasised in his report (made at the 1932 International Congress), that for the *ohm* and the *ampere,* the 1893 Congress defined them as units of the CGS electromagnetic system; the ohm was no longer essentially the resistance of a specified column of mercury column as had been the legal ohm. It was only represented by the resistance offered to an invariable electric current by a column of mercury ; the same reasoning applied to the *international ampere* which was sufficiently represented for practical needs by the rate of deposition of a current.

Fig. 9 Kelvin, drawn by L. Ward in 1897 (National Portrait Gallery)

He also pointed out that the representation adopted was from the practical viewpoint at the time the best possible but was not considered as definitive by the authors of the international system.

Following this 1893 Chicago Congress, France proposed an International Convention for the purpose of preparing laws to be passed by the different countries to render the international system mandatory; and the relevant legislation was enacted by the USA (12.7.1894), Britain (23.8.1894), France (by a Decree of 25.4.1896 to facilitate subsequent modification), and Germany (1.6.1898).[17]

6.5 International Electrical Congress: Paris, 1900

The fifth International Electrotechnical Congress held in Paris in August 1900 was like its predecessors held in conjunction with an international exhibition. Eighteen months of preparatory work were necessary for the organization of the Congress, and of the more than 900 participants about half were French. Some 120 technical papers were presented which when the Proceedings were published filled two volumes of more than 800 pages.

The Chamber of Government Delegates attending the Congress christened the CGS units of magnetic flux and flux density under the names of *maxwell* and *gauss*, respectively.[18]

The President, Prof Mascart, in reviewing the progress of electrical science since the beginning of the century, mentioned Napoleon who, after witnessing the experiments of Volta at the Institut des Sciences, had expressed the opinion that this branch of physics would open up the way to major discoveries. The first task, Mascart said, had been to evolve the principles and laws of electric current, this was accomplished over a period of thirty years by the work of Oersted, Ampère and Faraday. At the same time, Davy's memorable experiments had shown the possibilities for chemistry, and Arago's discoveries on magnetisation by electric current had led to telegraphy. Progress in the application of electricity for motors and dynamos had been slower, since although problems had been resolved in their theoretical aspects, the practical solutions presented greater difficulties. The small laboratory apparatus and battery-supplied equipment had now given way to machines of all dimensions capable of functioning for the most difficult and also the heaviest applications.

The natural energy from the rivers and falls, known as *la houille blanche*, had now been harnessed for the benefit of mankind. There had been a similar revolution in transport applications and in the industrial production of aluminium and the rarer metals, including copper refining and various new materials, which was only a beginning of all the possibilities open.

New progress was daily to be seen not only in electric lighting but also with the telephone and in telegraphy. The same wire could be used for simultaneous transmissions of several dispatches in two-way communications. Impressive results had been apparent in medical applications including the mysterious properties of the rays emitted from Crookes' tubes.[19]

In this field of investigation much remained to be done and present concepts of the constitution of the human body would be greatly modified.

Prof Mascart said, when one day our grand-children come to assess our 19th century, they would be amazed at all the accomplishments of this era.

After reading out a letter from Lord Kelvin, who for health reasons, had not been able to attend the Congress, Prof Mascart referred to the American delegates who had wished to bring some modifications to the present system of units. He was pleased that they had seen fit to withdraw these proposals since any changes to the units, recognized everywhere as official, such as the ohm and the volt, would have raised insurmountable problems.[20]

6.6 International Electrical Congress: St Louis, USA, 1904

There had been no international electrical congresses on electrical units or standards since the Paris Congress of 1900. One reason was that in view of the extent to which such units and standards had been incorporated in national legislation since the Congress of 1893, there had been less possibilities open for action in this field.

On the other hand, the need still existed for international governmental activity in bringing the electrical standards of the different countries into line and in raising the level of precision of such standards.

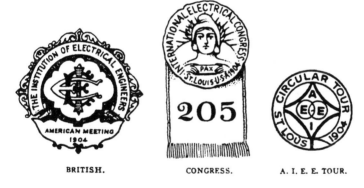

Fig. 10 St. Louis International Electrical Congress badges

However, it was considered that the various governmental bodies legally entrusted with the establishment and maintenance of electrical standards should participate in the work on a more continuous basis.

Furthermore, with the exception of the Congresses of 1881 and 1893, there had been criticism of certain members of the Chamber of Delegates whose scientific standing had not always been of the highest order. The editor of the *Electrical World and Engineer* shortly before the St Louis Congress questioned whether some delegates at congresses had been merely those who, intending to visit the Fair, had obtained their appointment as delegates 'with little or no reference to their qualifications for the duties involved'.

The International Electrical Congress of 1904 differed therefore from preceding Congresses in that its Chamber of Government delegates took no direct action in respect of units or standards. More than 700 participants, of which 150 came from abroad, attended the Congress at which about 160 papers were delivered on various aspects of electrical science and engineering.

The Chamber of Delegates approved the adoption of the following two resolutions: the first dealing with electrical units; and the second with international electrical standardization.

In view of the historical importance of these decisions, which marked a clear distinction between the two categories of future electrical standardization, the resolutions are given in full:

● Committee on International Electrical Units and Standards

It appears from papers laid before the International Electrical Congress and from the discussion that there are considerable discrepancies between the laws relating to electric units, or their interpretations, in the various countries represented, which, in the opinion of the Chamber, require consideration with a view to securing practical uniformity

Other questions bearing on nomenclature and the determinations of units and standards have also been raised, on which, in the opinion of the Chamber, it is desirable to have international agreement.

The Chamber of Delegates considers that these and similar questions could best be dealt with by an International Commission representing the governments concerned. Such a Commission might in the first instance be appointed by those countries in which legislation on electrical units have been adopted, and consist of (say) two members from each country.

Provision should be made for securing the adhesion of other countries prepared to adopt the conclusion of the Commission.

The Chamber of Delegates approves such a plan and requests its members to bring this report before their respective governments.

It is hoped that if the recommendations of the Chamber of Delegates be adopted by the governments represented, the Commission may eventually become a permanent one.

That the delegates report the resolution of the Chamber as to electrical units to their respective governments, and that they be invited to communicate with Dr S.W.Stratton (Bureau of Standards) Washington D.C.) and Dr R.T. Glazebrook (National Physical Laboratory, Bushby House, Richmond, Surrey, England), as to the results of their report, or as to other questions arising out of the resolution.

● Committee on International Standardization of Electrical Apparatus and Machinery.

The Committee of the Chamber of Delegates on the Standardization of Machinery begs to report as follows:

'That steps should be taken to secure the cooperation of the technical societies of the world by the appointment of a representative commission to consider the question of the standardization of the Nomenclature and Ratings of Electrical Apparatus and Machinery.

If the above recommendation meets the approval of the Chamber of Delegates, it is suggested by your committee that much of the work could be accomplished by correspondence in the first instance, and by the appointment of a General Secretary to preserve the records and crystallize the points of disagreement, if any, which may arise between the methods in vogue in the different countries interested. It is hoped that if the recommendation of the Chamber of Delegates be adopted, the commission may eventually become a permanent one.

That the delegates report the resolution of the Chamber as to International Standardization to their respective technical societies, with the request that the societies take such action as may seem best to give effect to the resolution, and that the delegates be requested to communicate the result of such action to Col R.E.Crompton, Thriplands, Kensington Court, London, England, and to the President of the American Institute of Electrical Engineers, New York City.'[21]

The International Electrotechnical Commission (IEC), thus created, had its inaugural meeting in London on June 26, 1906 when Lord Kelvin was unanimously elected its first President, with Col.R.E.Crompton as Honorary Secretary. (The work of the IEC on units and symbols is outlined in Chapter 7 and its associated Note 14).

6.7 International Conference on Electrical Units: Berlin, 1905

It had been recommended by the Chamber of Delegates at the 1904 Congress that an international conference on fundamental electrical units should be held because of 'discrepancies between the laws relating to electrical units and their interpretation'.

This international conference was duly held at Charlottenberg (Berlin) beginning on October 23, 1905. The main subject of discussion was whether the three units: *ohm*, *ampere* and *volt* (being connected by Ohm's law) should be independent of each other, or if only two should be defined, and if so which two. Dr Glazebrook (England) stated that the British Association had concluded that only two units, the *ohm* and the *ampere*, separately defined, should be selected.

According to Dr A.E.Kennelly – who refers to differences of opinion at the conference on whether the second primary standard should be the *volt* or the *ampere* – the British position had been defined at a meeting of the British Electrical Standards Committee only one week before the Berlin conference.[22]

Henry S. Carhart, professor of physics at the University of Michigan, presented at the conference a long paper (not reproduced here) setting out the reasons why the *ohm* and the *volt* should be selected as the independent units; the United States Bureau of Standards took a similar position in a paper also presented at the conference.

Meanwhile, Lord Kelvin had written to Prof Carhart on October 19, 1905, with a follow-up two days later, evidently too late to reach him in time for the meeting, agreeing that the time had come for making the *volt* and the *ohm* the two primary units, instead of the ampere and the *ohm* as was the case at present.

Kelvin's two unpublished draft letters setting out his reasons for this choice – which was at variance with the British viewpoint adopted by the conference – are reproduced below. It is surprizing that in spite of Kelvin's support for Prof Carhart, the United States recommendation was not accepted.[23]

The second letter to Prof Carhart dated October 21, contained additional information and deleted several passages of the earlier letter; these deletions are indicated by means of square brackets.

'Dear Prof. Carhart

I am very much interested in what you tell me in your letter of Sept 15 from Rhodesia regarding the meeting of the International Commission on electrical units and standards to be held at Berlin. I am glad to know, as I believe is the case, that all the nations to be represented are loyal to the recognition of Gauss and Weber's absolute system as our present foundation for all scientific and practical electric measurements and agreed to the C.G.S. detail.

The subordinate details for legal conditions regarding electric measurements or the supply of electric power (light included) and for electric pressure, especially in respect to safety of users and of the general public, are clearly of supreme practical importance.

I quite agree with your view that the time has come for making the volt and the ohm [the] two legally primary units, instead of the ampére and ohm as at [present].

[When the system was legalised in England, the electrolytic] verification of the ampere written 1/10th percent was surer and more easily practised by the silver voltameter or by the copper voltameter with Thomas Gray's working conditions than was a correspondingly accurate verification of the volt by a standard cell. But the progress of experimental research on standard cells has in my opinion wholly turned the table in this respect.

The direct verification of the volt by means of a standard cell is, at present I believe, considerably more accurate in common practice than the best present electrolytic representation of the ampére: and is certainly much easier.

Wishing you a pleasant time in Berlin, and begging that you will give my kind regards to all my scientific friends in this conference.'

Kelvin's second letter reads:

21st Oct 1905

'Dear Prof Carhart,

I wish to add to my letter to you of Oct 19 that my preference of verification of volt by standard cell to ampere by electrolysis is merely in respect to a subordinate method which should be admitted as a legal test.

I never for a moment intended to make it take the place of the absolute determination by Rayleigh's absolute electrodynamic balancer, on which the practical ampere is founded: and from which the ohm [and] the practical volt is fundamentally determined. In case of any question will you make this clear from me to the Conference, and explain that by legally I meant not scientific law, but practical law as to what is admissible and convenient in respect to supply and demand.

To make this quite clear please delete the word "the" before the word "two" in line 6 from the foot of the first page of my letter, and delete the last word of this line and the first 9 words of following line.'

The final conclusions of the conference were as follows:

1) That only two electrical units shall be chosen as fundamental units.
2) The international ohm, defined by the resistance of a column of mercury, and the international ampere, defined by the deposition of silver are to be taken as the fundamental electrical units.
3) The international volt is that electromotive force which produces an electric current of one international ampere in a conductor whose resistance is one international ohm.

In addition, it was decided that 'The Western Cadmium Cell shall be adopted as the Standard Cell'. Detailed specifications were recommended for the construction of the mercury ohm standards and of the Western Standard Cells.

The conference further resolved that since the laws of the different countries in relation to electrical units were not in agreement an international conference should be summoned in the course of a year in order to arrive at agreement in the electric standards in use.[24]

6.8 International Conference on Electrical Units and Standards: London, 1908

The British Association meeting in Dublin to hear a Report of the Committee on Experiments for improving the construction of practical standards for electrical measurements, and associated matters, expressed their deep sense of the loss they had sustained in the death of Lord Kelvin. It was recorded that the name of Kelvin, who had 'continued his active interest in their work up to the end', would 'always be associated with the establishment of the absolute system of electrical measurements and with the determination of the absolute units'.

The Committee expressed the hope that the international conference to be held in London would 'settle in a definite manner the few matters relating to the fundamental units which are still outstanding, and will organise a method whereby a close agreement may be maintained among the electrical standards in use throughout the world'.[25]

The Conference which subsequently met in London in October 1908 was attended by the representatives of 24 countries.

According to Dr A.E.Kennelly, 'the international electrical units were defined as at previous conferences in terms of the CGS units, but their standards, for legislation purposes, were recommended to be constructed according to specifications'.

The fundamental standards were to be the mercury-column ohm and silver-voltameter ampere, with numerical specifications. This controversial question regarding the adoption of the ohm and ampere, which had dragged on since the Berlin Conference, was decided by a vote of 19 countries in favour with four opposed.

An international committee was appointed to draw up improved specifications for the silver voltameter and to decide upon the emf for the Weston cell.[26]

According to the International Electrotechnical Commission, the government representatives adopted at the conference:

   a)   a set of fundamental units defined as decimal multiples of the corresponding CGS units,
   b)   a set of international units forming a system of units to represent the fundamental units and sufficiently close to them to serve for purposes of measurement.

These international units had been based on an international ohm and an international ampere defined in terms of the deposition of silver by an electric current.[27]

6.9 International Electrical Congress: Turin, 1911

A technical committee composed of representatives of the national laboratories of France, Germany, Great Britain, and the USA met at the Bureau of Standards in Washington in spring 1910 to make a comparison between the standards for resistance, current and voltage in use in the respective countries and reached agreement on the values to be adopted for international units.

The following year, in conjunction with the meeting of the International Electrotechnical Commission, the associated International Electrical Congress met at Turin.

This 1911 International Congress was in fact the last to be held as the Congress which was later proposed to be held at San Francisco in 1915, also in conjunction with a meeting of the IEC, had to be cancelled due to the outbreak of the first World War.

The Congress adopted a short list of algebraic symbols for the designation of certain electric and magnetic quantities. The most important decision was the international standardization of the symbology for ohm's law, which had differed in a number of countries.[28]

Notes

(1) *Life*, op cit, p 774
NOTE
Just prior to the Paris Congress, W.H.Preece, whose membership of the B.A.Committee dated from that year, submitted a long paper summarizing the electrical unit problem. He regretted the use of the term 'absolute unit' (first used by Gauss in 1832 in a paper *Intensitas vis magneticae terristris in mensuram revocata*, to indicate that the system was independent of any measuring instrument or gravity). In preference to this unfortunate term, Preece would have preferred the word 'invariable' or 'dynamic'. He stressed that the B.A. CGS system had made the submarine telegraph cable both technically and commercially possible, and that the results obtained from measuring instruments now used in research and inspection had superseded the quality measurements of the past. The progress particularly in lighting and power was directly related to these new means of arriving at an accurate relationship between the force exerted and output.
(W.H.Preece,Sur la mesure pratique des grandeurs électriques, *L'Electricien*, 1.10.1881, pp 545–564)
(2) P.Janet, *La vie et les oeuvres de E.Mascart*, Paris, 1910, pp 32–39
Bulletin de la Société Internationale des électriciens, 1908, p 640 (reproduced by G.Szarvady, *Unités électriques*, Paris,1919, pp 91,92).
Journal Télégraphique, 1881, Vol 5, pp 197–205.
(3) *Comptes rendus des travaux, Congrès international des électriciens* 1881, Paris, 1882.
(4) Michael Wooley, The International Congress 1881, *Telecommunication Journal*, Vol 48, Sept 1981, pp 543–547
(5) Sir William Thomson, Electrical units of measurement, *Popular lectures and addresses*, London, 1891, pp 93–96
(6) Resolutions 29.4.1884 and 2.5.1884 (Author's translation) M.E.Hospitalier, Congrés international d'électricité, Paris, 1900, *Rapports et procès-verbaux*, Paris, 1901, pp 15–17
Comptes rendus du Congrès international de l'électricité, Vol III, Paris 1932, p 533
(7) *Life*, op cit, pp 887 – 889.
(8) E.Hospitalier, *Congrès international d'électricité*, Paris,1901,pp 18,19
NOTE
Hospitalier complained, in *L'Electricien* 29.3.1890, that after a delay of seven months the *Minutes* had been received, with all the commas inserted at random, by the Government printing office.
(9) A.E.Kennelly, Hist.outline elect.units, op cit, p 246
(10) *Procès-verbaux sommaires, Congrès international des électriciens*, Paris 24–31 août 1889, Paris, 1890
(11) *Life*, op cit, p 889

(12) *L'Electricien*, 9.8..90, p 750
   NOTE
   An amusing report in *L'Electricien*, 6.9.90, pp 828- 823, signed Hospitalier(the editor), under the heading *HEN, GILB ET FRANK*, ridiculed an article by Prof J.A.Fleming (University College London) who proposed baptising several new electrical units after Henry, Gilbert and Franklin, names apparently overlooked by previous Congresses. The *HEN* would replace the *quadrant* for which it would constitute a fourth variant ! Although *HEN* had the merit of being short, a Frenchman, however relaxed, might give the impression that he was suffering from a cold.

(13) M.E.Hospitalier, *Rapports et procès-verbaux,Congrès international de l'électricité*, Paris, 18–25 August 1900, p 21
   NOTE
   The unofficial International Congress held at Edinburgh in 1892 was to consider the material specifications for the Chicago Congress 1893. (*Comptes rendus du Congrès international de l'électricité*, Paris,1932, Louis Joly, Report No 1 (France), L'état actuel de la question des unités électriques et magnétiques, p 534).

(14) *Measures for progress*, op cit, pp 31,32

(15) A.E.Kennelly, Historical outline of the electrical units, *Journal of engineering education*, Vol XIX No 3, Nov 1928, Pub of Harvard Engineering School No 30 (Reprint), pp 247–249

(16) Dr A.E.Kennelly, A historical sketch of International Electrotechnical Congresses, *Electrical World and Engineer*, pp 467,468

(17) Louis Joly, op cit,pp 534,535

(18) Kennelly, *Elect World*,op cit,p 468

(19) NOTE
   In his letter of Feb 13, 1892, Kelvin had written to Lord Raleigh concerning the discoveries of Crookes which he considered to be 'the greatest things by far, in deep-going physics of the nineteenth century'. (*Life*,op cit, pp 910,911)

(20) M.E.Hospitalier, *Rapports et procès verbaux, Congrès international d'électricité*, Paris, August 1900,pp 1–23, 346–370
   NOTE
   According to Dr A.E.Kennelly (*Hist.Outline.*,op cit, p 251),the American delegates had suggested 'fixing the working units of the magnetic circuit with names, in the CGS system...the gilbert for mmf, the oersted for reluctance, the maxwell for flux, and the gauss for flux density'.

(21) *Electrical World and Engineer*, Vol XLIV,No 13, September 24, 1904, pp 500– 502

(22) A.E.Kennelly, *Hist. outline.*,op cit, pp 254,255

(23) Kelvin Collection, Letter book: LB 28.66; LB 28.75, Cambridge University Library.
   NOTE
   Kelvin's incorrect use of the acute instead of the grave accent in *ampere* is curious, since it was he who proposed that the unit should be spelt without an accent. His second letter omits the accent.(*Life*, p 901).

(24) *Electrical World*, Vol XLVII, No 12, 24.3.1906, pp 614–616
   A.E.Kennelly, *Hist.Outline*, op cit, pp 253–255

(25) *Reports on the state of science*, British Association, 1908,pp 31,32

(26) A.E.Kennelly, *Hist.Outline.*,op cit, p 255

(27) IEC Publication 164: *Recommendations in the field of quantities and units used in electricity;* Section One, The history of electric and magnetic quantities and units, Geneva,1964, pp 20, 21
Rexmond C.Cochrane, *Measures for progress,*Washington,1966, pp 60,61: describes Prof Carhart's influence in the creation of the U.S.Bureau of Standards in 1901.
(28) A.E.Kennelly, *Hist.Outline,* op cit, pp 255–257

Chapter 7
Units and standards for the electrical century

7.1 Board of Trade Committee on Denomination of Standards

Britain had not been represented at the Paris International Electrical Conference whose decision of May 1884 had ruled that the legal *ohm* was to be a column of mercury of length 106 cm and section 1 mm^2 at a temperature of melting ice. This recommendation had, however, been communicated to the relevant Governments in order to obtain international recognition of this unit.

With the increasing use of electricity for lighting purposes, and the conclusions reached by the Paris Conference having proved open to comment, the Board of Trade had doubts 'whether the decisions arrived at in France should or should not be binding in this country'. In consequence, the Board wrote to the Royal Society on November 21, 1890 asking it to appoint one or two representatives to join a Committee to consider the question under the terms of the 1889 *Weights and Measures Act*.[1]

It was Thomson who, only ten days later was formally elected President of the Royal Society, presumably appointed himself and Lord Raleigh as members of the Committee. The Committee was set up the following month, and of the ten members representing the Board of Trade, General Post Office, British Association, Institution of Electrical Engineers and the Royal Society, seven were Fellows of the Royal Society. Its terms of reference were to decide what action, if any, should be taken 'with a view of causing new denominations of Standards for the measurement of electricity for use for trade to be made and duly verified.'

The extremely detailed Report of the evidence given to the Committee, published as a 95-page document in 1891, under the title *Standards for the Measurement of Electricity for use for Trade*, included an index, compiled in question and answer form, with the names of the expert witnesses called.[2]

As an example of Thomson's participation in the work of the Committee, Thomson stated (page 2) that no makers were so much affected as the British Post Office in respect of the ohm, and added, 'I am sure the Post Office has a hundred times as many standards of resistance as any 10 practical engineers in this country. I am sure I am not overstating it'. Later (page 8), he insisted that 'ampere' should be printed without an accent, 'because it is an English word now'.

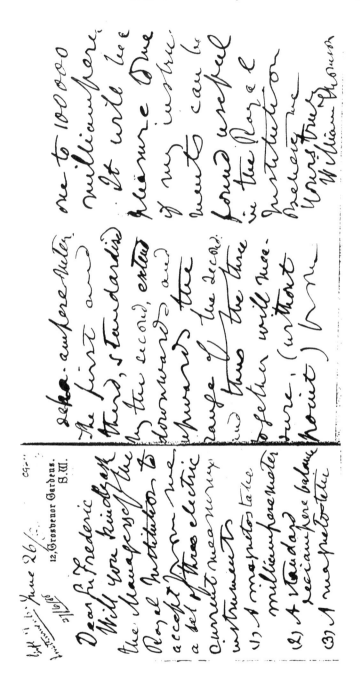

Fig. 11 Letter from Kelvin to Sir Frederic Bramwell, presenting the Royal Institution with a set of three 'standard' electric current measuring instruments

Thomson's commentaries and those of Lord Raleigh, Prof Silvanus Thompson and other scientists, as well as the evidence of practising engineers such as R.E.Crompton, etc, formed valuable contributions to the work of the Committee, and incidentally constitute an accurate record of the state of the art in Britain at the end of the 19th century.

The final resolutions of the Committee, based on the system of electrical units originally defined by the British Association and the work of the International Congress on Electrical Units in Paris, was finally published in July 1891. It related to the new denominations of standards for the measurement of electricity including those for resistance, the *ohm*, the *ampere* and the *volt*, and the various metal standards to represent them as well as to the associated measuring instruments.

It was subsequently ruled that for reasons of practical importance it was undesirable to adopt a mercurial standard for the measurement of resistance but that a material standard constructed in solid metal should be used.

However, this ruling became the subject of a *Supplementary Report of the Electrical Standards Committee*, which included Thomson, to the Board of Trade on November 29,1892, when the column of mercury method was accepted.[3]

This report explained that as the German Government had intended to establish legal standards, and to ensure complete agreement between the proposed units in the two countries, they had postponed action until after the visit of Professor Von Helmholtz who was attending a meeting of the British Association at Edinburgh in August.

There had been a full discussion of the subject in Edinburgh in which Dr Guillaume of the Bureau International des Poids et Mesures and Dr Carhart of the University of Michigan had also participated. With a few comparatively slight modifications the resolutions in the previous Board of Trade Report had been adopted.

This was to the effect that new denominations of standards for the measurement of electricity, whose magnitude should be determined on the basis of the CGS electromagnetic system of measurement, should be made and legally adopted as Board of Trade standards. A column of mercury was to be adopted as the practical unit of resistance as well as a standard of solid metal having the same resistance. These two standards were to be compared and verified periodically. The final clause of the *Supplementary Report* read:

> '16. That instruments constructed on the principle of Lord Kelvin's Quadrant Electrometer used idiostatically, and, for high pressures, instruments on the principle of the balance, electrostatic forces being balanced against a known weight, should be adopted as Board of Trade standards for the measurement of pressure, whether unvarying or alternating.'

It had been Kelvin (raised to the peerage in February) behind the scenes, who had personally arranged for the visit of Helmholtz to the B.A. meeting in Edinburgh. In his letter of June 20, 1892 Kelvin had explained that an Order was being issued defining for practical purposes standards of

Fig. 12   Kelvin's electrostatic voltmeter (Science Museum)

electrical resistance, current and 'pressure', and he added, 'I was asked by my colleagues to do what I could to persuade you to come yourself. You know I would compel you if I could !'

Later, on August 23, Kelvin was able to inform Lord Raleigh 'We had a splendid time in Edinburgh. Helmholtz...was most satisfactory as to units'.[4]

7.2 Kelvin rejects the 'Kelvin'

The Board of Trade had proposed to Kelvin early in 1892 that the name 'kelvin' should be given to the new Board of Trade unit of supply ('the energy contained in a current of 1 000 amperes flowing under an electromotive force of one volt during one hour') in place of the 'kilowatt-hour'.

In his letter to Sir Courtney Boyle of May 6, 1892 , Lord Kelvin rejected the proposal in favour of the term 'supply unit', on the grounds that people would not understand that he had no connection with various makes of supply-meters, or 'kelvin-meters' as they would become known, the more so if he should produce one himself:

'Dear Mr Courtney Boyle,

After further reflection, and careful consideration of circum-stances connected with the use of the name 'kelvin', should it be given to the Board of Trade Unit, I see objections which did not occur to me when I wrote to you two days ago, and which render the objection I had pointed out in my previous letter more serious than it first seemed to me.

There would be a great difficulty for the public to understand that Lord Kelvin has no part in instruments or inventions such as Aron's, or Teagues, or Schallenberger's, or Ferranti's supply meters, all giving their readings in 'kelvins', and if I succeed in producing a supply meter myself which may come into practical use, people might naturally be led to imagine that it only is genuine, and that the others are infringements of some priority, or right, that I might be supposed to have in respect to the measurement of supply.

I think you will agree with me that this is a strong objection against the proposed name, and will kindly excuse my having put it before you only after you have already had so much trouble which might have been spared had I pointed it out three days ago.

P.S. Instead of the somewhat cumbrous description 'Board of Trade Unit', the simpler expression 'supply unit' may readily come to be used, and will be quite convenient and satisfactory for signifying the particular unit defined by the Board of Trade for electric supply. It will seem quite natural that the readings of 'supply meters' should be given in 'supply units': and the ordinary householder will be quite satisfied with the simple word 'units' in checking and paying his electric lighting accounts. He, and the general public, know nothing of other electric units: and there is at present no other subject of reckoning in ordinary affairs to which the name unit is given'.[5]

However, the proposal for Kelvin's name to be assigned to the supply unit came up again barely a week after his death on December 17, 1907. A Board of Trade Memorandum dated December 23, initialled by Alexander P. Trotter, set out the reasons for assigning Kelvin's name to the supply unit.[6]

Firstly, the abbreviation, the B.T.U., as it would become known, could be confused with that for the *British Thermal Unit.* Secondly, 'the Board of Trade following a well-recognised custom in scientific and particularly in electrical nomenclature, proposed to adopt the name Kelvin for the 'Board of Trade Unit', and steps were taken, with the approval of the President and the acquiescence of Lord Kelvin. to insert in the Electric Lighting Provisional Orders of that year after the word 'unit', the words 'hereinafter called a *Kelvin*'.

As editor of *The Electrician* at the time, Trotter had been notified of the above 'kelvin' decision by Sir Courtney Boyle, who had undertaken to supply him with related correspondence for publication.

Trotter, who must have been somewhat embarrassed at the subsequent change of decision to adopt the 'kelvin', went on to explain:

'I at once adopted the word in *The Electrician* for May 6, 1892, but a week later, May 13, 1892, I published a letter from Lord Kelvin, saying that he had pointed out some reasons why the word should not be adopted, and he added that he was permitted to say that the Provisional Order would not introduce any new name.

I have reason to believe that the reason for Lord Kelvin's withdrawal of his acquiescence was that he was engaged at that time in inventing a meter, and thought it invidious that his meter should as it were set the example, and that all meters should register in Kelvins.

That reason no longer holds, and I beg to recommend that the steps which were taken in May 1892 be proceeded with.

The word *Kelvin* is fit for international use like the *Ampere, Volt, Ohm, Watt, Henry, Farad, Coulomb;* while the *Board of Trade Unit* is impossible for such use.

Sir Francis Hopwood tells me that he remembers the correspondence, and is sure that modesty was the only reason for Lord Kelvin's objection. Sir Oliver Lodge tells me that he remembers Lord Kelvin's objection to such a use of a man's name during his lifetime. I have spoken to several scientific men about this matter, and they all approve'.

Sir Alex Kennedy (of Kennedy and Jenkin), in his reply of April 8, 1908, to Trotter, returned a copy of a letter from Professor E.Mascart, Kelvin's French colleague at the series of meetings of the International Electrical Congresses and standards conferences, and who as President-elect, had he lived, would have succeeded Kelvin as first President of the International Electrotechnical Commission.[7]

The only objection Kennedy had to the name 'Kelvin' was the 'extreme awkwardness of the word itself'. Apart from this he saw nothing in Prof Mascart's letter which might hinder the Board of Trade from doing what apparently they long ago had determined to do. Prof Mascart, whose letter has not survived, had felt that the term itself would become 'neglected or useless.'

Professor Mascart, and Lord Kelvin himself, opposed the use of the 'kelvin' for the Board of Trade Supply Unit, perhaps in anticipation of something more significant. The name *kelvin* chosen – albeit some fifty years after his death – as the unit for thermodynamic temperature which was adopted as one of the units of the *International System of Units (S.I.)* now commemorates his name for all time.

7.3 MKS system: Giorgi's MKSA system : International System of Units (S.I.)

The first two decades of the twentieth century witnessed a new industrial revolution – the electrification of industry. In the USA in 1899, less than 5% of all power used in industry had been electric but by 1909, with the development of more efficient generators, electric motors, and power lines of greater capacity, it had risen to about 25%, and by 1919 had again increased to 55%.

In 1903 it was generally believed that it was not possible to make absolute electrical measurements to a higher accuracy than one in one thousand; by 1910, however, such measurements had been made with an accuracy of a few parts in 100 000, by which time a new era of high accuracy in electrical measurements had begun.

Defining an electrical unit was one thing, but determining its value relative to absolute units – that is, the same units used to measure mechanical energy, the centimetre, gram and second (CGS) – was another matter.[8]

At the International Electrical Congress held at St Louis, USA, in 1904, several papers on electrical theory had been presented. On September 12, Prof M.Ascoli in his paper 'Systems of electric Units' advocated the adoption of a rational system of units as proposed by Professor Giovani Giorgi (Italy).

The problem was also underlined by Dr F.A.Wolff, whose paper 'The so-called international electrical units', discussed the differences in the units as legalised by the principal European and American nations as well as the multiplicity of the same units in several governments. He urged that steps be taken to bring about closer unification.[9]

A letter from Giorgi (1871–1950), had been published in *The Electrician* in 1895 criticising the CGS system of units. In 1901 he had presented, at a meeting of the Italian Electrical Engineering Society, a paper *Unità razionali di elettromagnetismo*, in which he proposed a measuring system based on the metre, the kilogram, the second (MKS) which would be more convenient with the practical units of electricity, and that an electrical unit should be added to the three base units.

Giorgi was thus the creator of the first rationalized system, the MKSA or Giorgi system. The later choice of the ampere as base unit made it possible for the system to become coherent both for electrical and magnetic units, with the advantage that no conversion factors were necessary.[10, 11]

In 1904, Prof. Robertson of Bristol. who was in favour of Giorgi's system proposed the name *newton* for the unit of force.[12] Unfortunately, Georgi's proposed new system of units, neglected for more than thirty years, remained a purely academic exercise.

Prof Giorgi, however, relaunched his system in a *Memorandum on the MKS system of practical units*. This historically important 20-page text, printed in bilingual French/English form in June 1934 by the International Electrotechnical Committee (IEC) became a key document in international consideration of the new system. The foreword to the publication, added by the IEC, pointed out that the purpose of Prof Giorgi's *Memorandum* was to describe his proposal and to set out the principles on which the system was founded. The IEC text announced that:

> 'The Committee for Electric and Magnetic Units [of the IEC] at its meeting in Paris, October 5th and 6th 1933, voted unanimously in favour of the proposal to arrange the system of practical electrotechnical units into a complete absolute system (MKS system) which is intended to remain in use simultaneously with the CGS systems, in accordance with the suggestion advanced by the writer [Giorgi] in 1901.'

Giorgi's *Memorandum* emphasised that there were three groups of units in use: CGS electrostatic, CGS electromagnetic and the practical units. It had become necessary to adopt this third series of so-called 'practical units'; one reason was that the derived units of the CGS systems, even the mechanical ones, came out in very inconvenient sizes, originally defined as multiples of the CGS units. Giorgi explained:

> 'The CGS system had originated at a time when the whole theory of electric magnitudes and dimensions had not been perfectly developed. An 'absolute' system was then regarded as a system where all units are derived from three fundamental units, those of length, mass and time. The first aim in building the CGS system was to have a universal system for all purposes.'

He pointed out that in 1901 it had occurred to him that the solution to all difficulties accumulated up to then, could at once be obtained if the whole of the electrotechnical practical units were taken together with the metre as unit of length, the kilogram as unit of mass and the second as a unit of time, so that by adding one other arbitrary unit, a complete system of units of absolute character, was built up from four fundamentals.

After examining the details of his proposal, Giorgi concluded that his system would result in a great simplification of all practical calculations and of the learning of theory in schools, as well as a saving in time and intellectual effort. He left open the question of the choice and standardization of the fourth fundamental unit for future discussion.[13]

In 1935, the International Electrotechnical Commission unanimously adopted the Giorgi system of units. In an unpublished letter of June 27,

1935, Prof A.E.Kennelly, International Electrotechnical Commission, announced this decision to the International Union of Pure and Applied Physics:

> 'I beg to inform you that at its recent plenary meeting in Holland and Belgium, the Commission Electrotechnique Internationale adopted without opposition the GIORGI system of absolute practical units, commonly described in the mechanical order, as the M.K.S. system. It was recognized that, in order to complete the system formally, four fundamental units are necessary. It is agreed that the fourth fundamental unit should be selected from among the practical units already adopted, i.e. among the OHM, VOLT, AMPERE, COULOMB, FARAD, HENRY, and WEBER. The WEBER is the new unit of magnetic flux in the practical system of electromagnetic units adopted by the IEC in the recent Holland and Belgium plenary meeting.
>
> It was decided that the choice should not be made before consulting the International Union of Pure and Applied Physics, S.U.N. Committee; and the Comité International des Poids et Mesures, Comité Consultatif d'Electricité.
>
> Professor L.Lombardi was requested to communicate with the Comité Consultatif d'Electricité, and I was to communicate with the S.U.N. Committee.
>
> I therefore beg to bespeak the kind attention of the S.U.N. Committee of the Union internationale de physique in the hope of obtaining, if possible, its valued opinion upon this question, which is manifestly of importance both to Physicists and Electrotechnicians; because it would evidently be most regrettable if the choice and definition of the fourth fundamental and practical unit should lead in the future to incomplete agreement between the practical units used in Physics and Electrotechnics.'

A long 'confidental' note by Sir Richard Glazebrook, under the heading 'Proposal of the International Electrotechnical Commission to adopt the Giorgi system of electrical units', was forwarded at the same time to the British National Committee of the IEC, calling for the views of members of the Electrical Units and Standards Committee of the National Physical Laboratory.

Glazebrook emphasized 'that the MKS system and the Giorgi system were not the same in that although both systems adopted a fourth unit, the MKS system left that choice of unit open, while the Giorgi system called for 'some material standard , e.g. the International Ohm defined as the resistance of a certain column of mercury, or the International Ampere defined in terms of the electrochemical equivalent of silver or some other unit...'[14]

As a result the CGS system, which had been adopted at the 1881 International Electrical Congress in Paris, was duly superseded, the fourth unit the ampere being adopted in 1950.[15]

The metre, kilogram, second, ampere (MKSA) system as it became known in turn paved the way for the internationally adopted practical system of measurement units, the *Système international d'unités (S.I.)*, which name, and rules for the prefixes, the derived units and the supplementary units was adopted by the Conférence Générale des Poids et Mesures (CGPM) at its 11th General Conference at Paris in October 1960.[16]

Notes

(1)   Letters 21.11.1890 and 16.12.1890, Board of Trade to Royal Society. (R.S. MC 15.111 and MC 15.117).

(2)   *Report of the Committee appointed by the Board of Trade: Standards for the measurement of electricity*, London, 1891. (House of Lords Record Office)

(3)   *Supplementary Report of the Electrical Standards Committee*, 29.11.1892. (Public Record Office, BT101/375).

(4)   *Life*, op cit, pp 923,925

(5)   Letter 6.5.1892, Kelvin to Sir Courtney Boyle, Royal Society. *Life*, op cit, pp 917,918

(6)   Letter 8.4.1908, and enclosure (Public Record Office, BT 101/392).

(7)   NOTE
      As representative of France, Prof. Mascart, had attended seven International Conferences including the great Congress of 1881. He organized and presided over the Second International Electrical Congress to reach decisions on the definitions of electrical units to form the basis of national legislation held at Paris in 1889 – postponed in the two preceding years at the request of the French Government as a French Commission was still investigating the question of electrical units and standards. The Congress adopted the units and unit names of the joule and the watt in terms of the C.G.S. units and also in terms of the practical system.

(8)   *Measures for progress*, op cit, pp 103–105

(9)   *Electrical World and Engineer*, Vol XLIV, No 13, 24.9.1904

(10) *DSB*

(11) NOTE
      At her *leçon d'adieu* (to which the Author was invited) at the Ecole Polytechnique Fédérale de Lausanne (EPFL), Prof Hamburger recalled the brain-splitting exercises of earlier days in passing from the CGS electrostatic system to the CGS electromagnetic system, and vice-versa. With coefficients and powers of 10 to be taken into account, she said, you had to think hard to remember whether they went in the numerator or denominator of an expression. (Prof Hamburger, Les unités de mesure dans l'optique de l'ingénieur, *Bulletin de l'Association Suisse des Electriciens*, 9.2.1980)

(12) B,Swindells, Understanding units of force, *Engineering*, Feb 1971, p 770

(13) G.Giorgi, MIEE, *Memorandum on the MKS system of practical units*, International Electrotechnical Commission (Advisory Committee No 1 [subsequently Technical Committee No 1] Section B: Electric and magnetic magnitudes and units, London, June 1934, pp 3,5,18
      NOTE
      The adoption by the IEC of Prof Giorgi's proposals was reported in *Nature* (6.7.1935) under the heading 'Electrical Units and the I.E.C.' the proposals had already been discussed in *Nature* the previous year (21.4.1934).

(14) Unpublished letter (and proposal) 27.6.1935, Prof A.E.Kennelly, Chairman of EMMU Committee of the International Electrotechnical Commission to Prof Abraham, General Secretary, Union internationale de physique pure et appliquée. Unpublished letter 3.7.1935, Sir Richard Glazebrook to Mr Griffiths (Electrical Units and Standards Committee of the National Physical Laboratory),(Public Record Office, DSIR/10/20)
NOTE
On its creation in 1906, the scope of the IEC included Nomenclature, under Technical Committee No 1 whose work covered: terminology; electrical and magnetic quantities and units; symbols. In 1935, the work in each of these sections had increased to the point that the two latter sections were detached under two independent IEC committees:TC24 and TC25.
In the evolution of electrical units, it has only been possible to outline the role of the IEC and other international organizations, including the Bureau International des Poids et des Mesures, the International Union of Pure and Applied Physics, and ISO/TC 12 (for non-electric questions). For the contributions of IEC/TC24, later taken over by IEC/TC25: Quantities and units,and their letter symbols; the reader is referred to a paper by Prof E.Hamburger, Quelques expériences à la tête de commissions techniques dites horizontales, *Bulletin de l'Association Suisse des Electriciens*, 6.6.1981, pp 578- 582; and IEC Publication 164(1964):Recommendations in the field of quantities and units used in electricity; both of these sources cover the history of electrical units. The work of IEC/TC 1:Terminology, which has continued without interruption since 1906, has been examined by the Author: Creating the International Electrotechnical Vocabulary, *Multilingua* 2–1 (1983), pp 35–37. (See App D).
(15) H.Moreau, *Le système métrique*, Paris, 1975, p 77
(16) Bureau Internationale des Poids et Mesures, *Le Système International d'Unités*, 4th ed., Paris, 1981, pp 5, 29.

Chapter 8

Kelvin, Crompton and the electrical industry

Colonel R.E.B.Crompton CB,FRS (1845–1940), a close friend of Kelvin, had the rare distinction of being an Honorary Member of the three senior engineering institutions in Britain.

Following military service, mainly in India where he did much to develop mechanical traction, in 1876 be became a practising engineer. An outstanding pioneer in the fast developing electrical industry particularly in the electric lighting sector, then in its infancy, as a designer and manufacturer of dynamos his fertile brain and boldness in introducing new designs brought him considerable success.[1]

His business acumen combined with his emphasis on quality at all levels, good working relationships, team-work and loyalty helped to make his factory an outstanding one even by today's total-quality standards in management and production. Crompton-trained electrical engineers were highly regarded in the profession.

He was twice President of the Institution of Electrical Engineers – first in 1895 and again in 1907 following the death of Lord Kelvin. The author of many papers in engineering science and numerous patents, Crompton was elected in 1923 to Honorary Membership and in 1926 was awarded the Institution's highest distinction, the Faraday Medal.

8.1 International standardization

Throughout his career he gave much attention to questions of standardization, the establishment of electrical units, accurate measurement and terminology .

His close relationship with Lord Kelvin extended over many years, and as will be seen below, on the founding of the International Electrotechnical Commission (IEC) in 1906, under the Presidency of Lord Kelvin, Crompton was elected Honorary Secretary. In 1931 he was elected Honorary President, and at the banquet on his 90th birthday he said he hoped to be remembered, more than anything else, for his work with the IEC.[2]

Whenever Crompton required guidance he did not hesitate to consult Kelvin. Both men had the ability to find original solutions to the numerous problems confronting electrical engineers. While the one depended solely on his engineering experience, the other brought to bear his considerable background of scientific investigation and mathematical reasoning, albeit tempered with a businesslike approach to electrical engineering.

8.2 Measuring instruments

With the rapid progress in heavy electrical engineering, a pressing need arose for reliable measuring instruments suitable for the relatively high currents used in lighting and transmission. As a result a number of ingenious and useful instruments were invented whose manufacture formed an important sector in the electrical industry.

In a paper read to the Society of Telegraph-Engineers and Electricians in 1884, Crompton described how he and his assistant G.Kapp, had designed what he termed current and potential indicators whose calibration would not alter as a result of external disturbing forces. He demonstrated the working of two different patterns of current indicators. The scale of such instruments was divided directly into amperes or volts which obviated the use of a calibration coefficient by which the deflection had to be multiplied. Hitherto such a calculation, involving decimal fractions, was in Crompton's view a tedious task for the unskilled workmen who had to use these instruments. Crompton's achievement in making instruments which did not require the use of constants or any table of values was thus of major practical importance to the electrical industry.

Measurement of current and 'potential' by direct comparison with a standard unit was replaced by an indirect method of measuring, not the current itself, but its chemical, mechanical, or magnetic effect.

While the chemical method was accurate, the time factor and the need for an absolutely constant current restricted its use to the laboratory or for the calibration of other instruments. For practical purposes, in the words of Crompton, 'we weigh, so to speak, the current against the force of a magnet, of a spring, or of gravity'.

Crompton's method was to excite the electromagnet by the current which was to be measured. In his paper he examined the various problems encountered when calibrating measuring instruments, for example, the elimination of the magnetic effect of the coils of the electromagnet on the indicator needle by introducing an opposite magnetic effect.

In the discussion, Dr J.H.Hopkinson related his 12 months' experience using three of Sir William Thomson's instruments for measuring both potential and current. The constancy of the magnets had been tested frequently and over about thirty trials no serious change had occurred. The objection to the use of such instruments and to the majority of galvanometers used for the measurement of current was the degree to which they were disturbed by magnetic forces in their vicinity. In the case of Sir William's instruments he had overcome this problem by altering the mode of suspension and using an astatic needle.[3]

In the electric measurement sector Crompton designed a number of instruments which provided accurate and reliable readings. Together with his assistant Gisbert Kapp, he patented several instruments, and in an interview in 1893, Crompton said he could measure voltage and current to an accuracy of 1 in 10,000.[4]

At a *Conversazione* held at the Royal Society in 1895, under the Presidency of Lord Kelvin, Crompton exhibited the latest form of Crompton potentiometer. On the same occasion, his platinum thermometers for use with a potentiometer were on display. The experimental apparatus showed the small consumption of electrical energy required to maintain a small crucible at a constant high temperature.[5]

In his lecture at the Royal United Services Institute in 1881, Crompton referred to Sir Humphrey Davy's spectacular arc produced at the Royal Institution in 1808. Lighting had remained a laboratory phenomenon with the galvanic battery 'expensive, fuming, uncertain and cumbrous' until Faraday's discoveries in dynamic electricity gave a fresh impulse to the subject.

Crompton pointed out that the great stride was made in 1870, when 'a clever French workman, Gramme, combining in his dynamo-electric machine the self-exciting or reaction principle, discovered by Wheatstone, Siemens, or Varley (it matters not which),with his own annular armature, gave us for the first time a really cheap electric current...' [6]

Crompton had immediately imported several sets of Gramme machinery and Serrin lamps and installed electric lighting in his Stanton works. When later in 1878 he went over to Paris with a party of mechanical engineers he found himself regarded as a leading authority on electricity. He then took the decision to go ahead and manufacture his own electrical plant.[7]

In 1880, Crompton visited Windsor Castle[8] to demonstrate the new electric lamp to Queen Victoria, who took a keen interest in the work. Following the installation at Buckingham Palace in 1884, the Queen expressed her dislike for the concealed lighting and asked Crompton to remove the lamps behind the cornices. After he had suggested she should wait and see, the Queen apparently punched him on the shoulder because, in his words he had had 'the cheek to have his own way'. The next day she agreed to everything.[9]

Crompton proudly informed Kelvin that the two machines (which Kelvin had seen at the Crompton works) delivered to the Palace required only 30 hp to do the work for which the Victoria Brush machines required 34 hp – a difference in efficiency of 12%.[10]

8.3 Crompton opposes metrication

Prior to the debate in the House of Lords in 1904 (referred to in Chapter 3), the question of metrication was discussed at length by the Institution of Electrical Engineers in January 1903, and again the following month.

Lord Kelvin's written communication emphasised 'that the universal adoption of the French metrical system by electrical engineers and engineers of all classes and common-sense people throughout the country, and in all saleshops and workshops and factories, will be a great blessing to every individual person concerned...' [11]

It is to be regretted that Crompton, one of the most vigorous supporters of standardization, spoke at length against the metric system. He questioned 'whether a change to the metric system was possible, or even desirable'. His main argument against metrication was the expressed objections to the system in the United States, 'reported in *Engineering* of January 23, 1903, pp 104 to 107', mainly on the grounds of the heavy costs involved in scrapping obsolete plant, and particularly in connection with screw threads.

He insisted on the superiority of the Whitworth system of threads over a decimal system, and considered that as most American and British engineers used a slide rule for their calculations there would be no loss of time involved in converting from one system to the other.

Kelvin must have shuddered on reading Crompton's comment that 'the tail cannot wag the dog, and in this particular instance the tail is the metric system and the dog is the inch'. The dog in fact had its day, and as has been already noted earlier, both the United States and Britain are still out of step with the rest of the world in regard to the full acceptance of the metric system.

In his reply, Alexander Siemens, who favoured the metric system, said that the main reason for advocating a change to the metric system was its convenience for international trade. In referring to the standardization of screw threads, he regretted that the conclusion arrived at in all the committees [the B.A. Committee had included Kelvin and Crompton] that had considered the question 'was that it was practically impossible to make screws fit which had been manufactured by two different manufacturers, unless standard screws and cutters were deposited in some place where everybody could go and compare his screws with them'.[12]

The discussion was fully reported in the leading newspapers. According to one technical journal:

> 'The preponderance of opinion was in favour of the abolition of the existing method; but unanimity that the metric system as it at present exists was the best was by no means evident. With one or two notable exceptions including Sir John Wolfe Barry and Colonel Crompton, all the participants in the debate were convinced that some change is needed; but, as Colonel Crompton remarked, a good deal of irrelevancy was indulged in and it being beyond the province of the Institution to pass a resolution of any description, the matter ends here.'[13]

8.4 Electric lighting and the 1881 Paris Exhibition

Crompton sent over to the Exposition d'Electricité at Paris in 1881 a large exhibit of generating equipment, switchboards, and arc lamps. Lectures at the exhibition covered such subjects as telegraphy, generators, submarine cables, railway signalling systems, recorders, clocks and electric motors, as well as electro-domestic applications. The lamps of a score-or-so manufacturers lighting up the skies of Paris attracted numerous visitors.[14]

Crompton's exhibits gained a gold medal, but in his view the most important result achieved was the standard electrical terminology decided at the International Conference.[15]

The first wide-scale application of incandescent lamp lighting was installed by Crompton at the Law Courts at the end of 1882. With a view to improving the installation, Crompton asked for Kelvin's advice.[16]

> 'Would you mind allowing me to have the use of your design for coupling in the accumulators to your dynamo circuit at your house, as I propose to try it at the Law Courts? ....I propose to subdivide the current into about 3 parts, two-thirds or say 600 amperes going to the batteries'.[17]

Kelvin replied by return of post advising Crompton that the existing instruments for this purpose could only carry about 50 amperes but that he could easily make one that would carry 300 or more:

> 'I think, for your purpose, a *part* of my plan will be sufficient, and will be simpler and less liable to any possibility of derangement than the whole. If therefore you wish it I shall instantly start making an instrument fit to carry 1,000 amperes, and adapted to break circuit before the current reverses; but without any appliance for re-making the circuit, if the potential at any time after becomes righted.
>
> As you will always have someone on watch, it seems to me that the automatic re-making of the current, which I used for my home lighting, with nobody on watch in my laboratory, will not be needed by you. The safety will be attained by the circuit being broken before the current goes the wrong way. When at any time it is desired to resume charging the accumulators, your man on watch will remake the circuit by hand. Tell me what you think of this and I shall go on as soon as I know what you would like to have.'[18]

Crompton placed great reliance on Kelvin's practical advice, and in a long letter written at the end of 1886, he informed him that he had been experimenting with other people's meters over a long period. He had used some of Ferranti's meters and obtained very fair results. He had also been working with the 'Compteur Couvray', a French invention, and intended to use both at Kensington Court.

As regards 'current and potential instruments', he had been improving and cheapening those made on the Crompton-Kapp patent, and were 'extremely accurate in fact to within 3%' provided they were not positioned in situations where they were affected by external magnetic fields; Crompton added:

> 'I am fully aware of the value of your remark as to the great influence the leads have on the accuracy of the measurements when strong currents are employed and am careful on this point. I have also used your clip and this is not such an easy matter when

very large currents are employed. It is necessary then to use a pair of flat plates of considerable depth separated by mica.'

Crompton closed his letter with a brief description of his electric lighting installation at the railway goods yards of the newly constructed Tilbury Docks:

'There were '1 300 incandescent lamps worked on the simple parallel system, the E.M.F. being 280 V at the terminals of the dynamos. The lamps are large, of 135 V placed two in series. I think this large installation will be worked exceedingly cheap, and is a fine example of large works of a simple character. The most distant incandescent lamps are 1 000 yards away from the generating station. The arc circuits are 3 or 4 miles long each.' [19]

8.5 The Dynamicals

This informal group of leading electrical engineers and scientists, including Crompton and Kelvin, was first proposed by W.D.Gooch, an electrical engineer, in March 1883. The members discussed matters of mutual interest while dining in friendly fashion in comfortable surroundings.

Crompton represented the electrical manufacturers while other members were associated with the railways, architecture, the Services, finance and even literature. At the invitation of Crompton, Kelvin became the Society's first president, a post he held for many years; other famous names included several members of the Royal Society such as Sir James Mackenzie, James Matheson, S.Z. de Ferranti, Prof. J.A.Fleming, Prof. George Forbes, Sir William Crookes, Prof.W.G.Adams, Sir W.H.Preece, Prof. Fleeming Jenkin.

In 1889, the society having abandoned the work of getting the *Electric Lighting Act* of 1882 amended, fell into decline. By then the Society of Telegraph Engineers, which had become the Institution of Electrical Engineers, was the representative body of the profession on technical and legislative subjects. But the Dynamicals, which by 1904 had become a purely dining and social club, is still in existence although it meets only two or three times a year. The original Minute books of the Society are still extant. [20]

Kelvin attended many of the monthly dinners, and on being re-elected President wrote: 'As you think my sadly incomplete performance of the duties could be excused I shall be happy to accept re-election to the office of president'.

For several years following the passing of the *Lighting Act* of 1882, which gave local authorities the right of purchase of private electrical supply companies after 21 years, investors became reluctant to entrust such companies with their savings. As a result, the development and extension of electric power installations in Britain became blocked.

This reacted in turn upon the electrical manufacturing industry, so that British manufacturers became unable to meet the requirements of either home or foreign markets. It was in such a climate that in 1885 the Dynamicals submitted a request to the Board of Trade to enable them to put forward their reclamations:

Electric power   *Jan. 21. 1885*

THE DYNAMICABLES*

Dear Sir

I have the honour to inform you that at the formal meeting of The Dynamicables on the 20th, it was resolved:

1. That Mr Chamberlain shall be asked to receive a Deputation from 'The Dynamicables' for the purpose of handing to him their Report on the Electric Lighting Set Committee

2. That Lord Bury, Chairman of the Electric Lighting Act Committee, be asked to introduce the Deputation to Mr Chamberlain

3. That the following gentlemen be asked to form the Deputation:

Sir William Thomson	*Prof. W.G. Adams*
Lord Bury	*Latimer Clark*
R.E. Webster	*R.E. Crompton*
R.H. Milward	*Musgrave Heaphy*
Prof. G. Forbes	*Dr. Hopkinson*
Arnold White	*Prof. Fleeming Jenkin*
Sir Charles Bright	*F.L. Rawson*
W.H. Massey	*Willoughby Smith*
F.R. Reeves	*C.E. Spagnoletti*
Dr. Stone	*F.H. Webb*

I shall be much obliged to hear from you at your early convenience whether you will act, as suggested, on the Deputation –

Yours truly

W.D. Gooch

Hon. Sec.

Sir William Thomson

---

*"Dynamicale": D126, Cambridge University Library

The delegation was received by the 'polite but somewhat sarcastic' Mr Chamberlain who listened to their demands for certain of the provisions of the *Lighting Act* to be regulated in the same way as the Gas Companies.[21]

But it was not until 1888 that an Amending Act extended the tenure of private companies from 21 to 42 years, which encouraged investors to put their money into such undertakings. As Crompton put it, 'during the years that followed the *Lighting Act* of 1889 the work of supplying London with electric light went rapidly forward'.[22]

At the Board of Trade enquiry set up in 1890 to report upon the standards in use for the measurement of electricity, Crompton's evidence as a practising engineer constituted a valuable contribution to the work of the Committee whose members included Lord Kelvin. (See also on this subject Chapter 7).

8.6 Crompton's 'Little Eye'

Dr Gisbert Kapp, who was head of Crompton's design office for some years, read a paper at the Institution of Electrical Engineers on Characteristics of dynamos. Crompton, present at the discussion on November 11, 1886, had this to say:

> '... As I do not intend to criticise this paper from a mathematical standpoint, I propose to say a few words for the comfort of those electrical engineers who, like myself, are not sufficiently skilled mathematicians to work out the design of a dynamo in the manner given in the present paper. There are, I dare say, many people present who, like myself, would never have designed a successful dynamo if it had depended on the abstruse and difficult mathematicial formulae which are now put before us by Mr Kapp. I, for one, have done most of my designing by the graphic method – that is to say, by eye. My practice has been to make designs and models, and introduce variations from time to time, constantly correcting these designs by eye...'

One week later Crompton, whose remarks had been publicly given, wrote to Kapp:

> Perhaps you don't know it but – (Air – 'I've got a little list' – Mikado) whenever I require to design a dynamo! [23]
> I've got a little eye! I've got a little eye! For whatever you may tell me, I assure you its 'no go'.
> With mathematics high! With mathematics high!
> I'm not one of these people who have got Clerk Maxwell pat.
> Those people who when 'Talking shop', talk shop to you like that.
> I'm not like those who always like to zero integrate
> The number of lines of force (or do I mean 'equate' ?)
> Nor can I work the calculus – nor do I mean to try.
> It's all my little eye, it's all my little eye.[24]

In his Presidential Address to the IEE in 1895, Crompton emphasised that he was the first IEE President whose early training in mechanical engineering had been entirely unconnected with electrical engineering. In his view, the great development in electrical engineering dated from the Paris Exhibition of 1878 when the mechanical engineer first took in hand the design and construction of dynamo electrical machinery.

Since that date the dynamo, the alternator, the transformer and the storage battery had been developed; the dynamo had increased in size by something like 500 times and its efficiency from 60 to 97%.

All this had been made possible by the pioneer work of the physicists and telegraph engineers who had laid down so clearly the laws and the co-relation of electromotive forces, electric currents, resistance and induction, and to the admirable system of units which through the British Association Committee had been brought to such completeness that they now had been universally adopted.

A major problem confronted designers in the relative strength of insulating materials which did not sufficiently resist temperature rises and were easily damaged by oil or grease. Manufacturers also experienced difficulties in cutting and shaping such materials.

During his lecture Crompton on several occasions used the term *load factor* [he had used the term slightly differently in 1891 in his ICE paper] which he had invented, and improved to denote the total useful time that an appliance was worked discontinuously over a certain period of time. In so doing, he drew attention to the periods of idleness of machines in a factory which largely exceeded their working time. His observations on load-factor, he emphasised, were of equal importance to the engineering world in general whether waterworks, gasworks, and many other industries. This term is now defined in the *International Electrotechnical Vocabulary* of the IEC.

Turning to the subject of accuracy of measuring instruments, Crompton referred to the claimed efficiencies of the various classes of electrical apparatus, which were worked out to several decimal places, when it was well known that no instruments could measure to such accuracies. He had spent the previous eight years [from 1877] in perfecting the potentiometer. With his instrument there were no moving parts and the only calibrated portion – the slide wire – could be subjected to extremely rough treatment without materially altering its accuracy.[25]

The editor of *The Electrician* in 1895 praised Crompton's 'little eye' which had helped England to take the lead in first-class engineering construction and design, and that 'in the widest and best sense of the term, there is no doubt that Mr Crompton is the leading electrical engineer in the land'.[26]

8.7 The National Physical Laboratory

As early as 1885, electrical engineers were demanding a national standardizing laboratory. Crompton was in the forefront of those who insisted that their measuring instruments should be backed by some kind of certificate attesting to their accuracy.

Fig. 13   Advertisement for Crompton and Co. Ltd (The Crompton Collection, Science Museum)

At the Institution of Electrical Engineers he urged that the need for such a standardizing laboratory was so great that if the Government did not take any steps in that direction, the electrical manufacturers would be obliged to combine and do the work themselves. He pleaded:

> 'We cannot go on any longer as we are doing now without independent check on our individual attempts at calibration. I have been at great pains and expense to get our own instruments correctly standardardized, but the results have not been entirely satisfactory.' [27]

Crompton made his own standards where these did not exist; in this way, he arranged for the conductivity of copper, then produced by electrolytic means, to be measured by using 'a few large accumulator cells for instantaneous measurements' and he rapidly became the testing centre for all the manufacturers of this class of copper.[28]

In 1889, Kelvin in a letter to his friend Mascart, informed him:

> 'I have been kept very much in London by duties as President of the Institute of Electrical Engineers, and I shall be detained still near the end of the present month, with official business and duties connected with the establishment of an electrical stan- dardizing laboratory, which we hope is to be taken in hand by our Government.'[29]

When the National Physical Laboratory was eventually founded in 1900, the staff of 25 worked in two divisions : physics and engineering. The first director appointed by the committee set up by the Royal Society was R.T.Glazebrook (afterwards Sir Richard) who was to work very closely with Crompton in several fields. Two years later, Crompton was appointed to the first Executive Board on which he continued to serve until 1923.[30]

At the official opening of the NPL on March 19, 1902, the Prince of Wales said that scientific knowledge would thus bear upon industrial and commer- cial life, break down barriers between theory and practice and effect a union between science and commerce.[31]

In seconding the Vote of Thanks to the Prince, Kelvin emphasised the importance of exceedingly minute and accurate measurements, which although they might not strike the popular imagination, were the founda- tion of the most brilliant discoveries.[32]

Crompton's period of membership of the Executive Committee and the General Board of the National Physical Laboratory continued unbroken from 1902 up to 1923. He had been a member of the Electricity Sub- Committee set up in 1899, and a member of the Tank Advisory Committee, the Buildings Committee, Laboratory Funds Committee, etc. He rendered much assistance in the work of the Metrology Department, especially in connection with the measurement and standardization of gauges.[33]

8.8 Electrical engineering problems 1905

In 1905, the Institution of Civil Engineers called upon Crompton to present a paper 'Unresolved problems in electrical engineering'. With his customary thoroughness, Crompton obtained the advice of Kelvin who advised him to stick to the practical aspects of electrical engineering. In view of its general interest this letter is reproduced in full:

> 'I have been overwhelmingly busy since I received your letter of Feb 13 or I should have sooner answered.
>
> I think your own views on solved and unsolved problems in Electrical Engineering will make an admirable subject for your James Forrest Lecture. I do not think it all requires very much reference to the 'new views' relation between mattter and form in electricity.
>
> As for electricity the new views are simply the adoption of atomic theory of electricity so far added to Aepinus 'fluid taking in every piece of finite matter, a finite number of electrical atoms. Helmholtz gave an atomic theory of two electric fluids: but there is now a very general concensus in favor of the older view of Aepinus; modified only by making the electric fluid be resinous electricity instead of vitreous electricity as Aepinus made it.
>
> You will find this somewhat fully explained in Appendix E of my Baltimore Lectures entitled Aepinus atomised, being a reprint from the Phil-Mag 1902 first half year. As for terrestrial magnetism, you will find nearly all I could say in my presidential address to the Society of Telegraph Engineers in 1874 reproduced in pp 206–238 Vol II of my P. of Inst. Address.
>
> I am sorry not to be able to be more helpful to you, but really I don't think you want help at all when you have such a subject given to you as 'Unsolved problems in electrical engineering' with perfect liberty to add a good deal of your own views on solved problems, which will certainly be most interesting to your readers.' [34]

Following Kelvin's advice, Crompton dealt primarily with problems that were of immediate concern to practical electrical engineers. These included:

- lightning discharges particularly where they affected distribution of large electric power plants;
- 'core and coil' problems in electrical machinery design;
- electrical measuring instruments including the power measurement of alternating current;
- electric smelting and the regulation of high temperatures over long periods in small furnaces.

Crompton considered that in electrical engineering, the sector in which there had been the greatest advances was measuring instruments. 'This is a matter on which I could dwell with loving interest,' said Crompton, 'but time prevents me from dwelling on it at any length'. He elaborated on this theme:

> 'The instruments that we have already at command for measuring continuous currents are already so accurate and so low priced that they are commencing to be used for many purposes quite

Fig. 14 Sketch of Crompton (The Crompton Collection, Science Museum)

outside the field of electrical engineering proper...but when we come to the measuring of alternating energy, although the instruments for measuring separately the current and pressures are satisfactory, yet no completely satisfactory method at all approaching the accuracy and simplicity of the continuous-current methods for measuring power is as yet available in this case.'

In Crompton's view, the problem of portable storage of electrical energy could not be resolved until they had what Lord Kelvin had first termed 'a box of power'. This had now been achieved as seen in the large number of electrical carriages on the streets of London and Paris. He said, with prophetical insight, that progress would be achieved mainly in the improved organization and standardization of the methods of using, and the facilities for re-charging and repairing the accumulators already in existence.

He emphasised the role of standardization in facilitating manufacture and interchangeability of component parts.[35]

8.9 Electrical standardization

The question of screw threads had been a problem for Crompton and other electrical engineers throughout his career. In 1901 he was asked to became a member of the British Engineering Standards Association to standardize small screws.[36]

Crompton was largely instrumental in designing the BSF (British Standard Fine) system of threads for which the first screws and taps were cut in Crompton's laboratory. This thread was designed to replace the standard Whitworth thread which was too coarse for electrical engineering applications. During World War I, Crompton served on a committee in the Ministry of Munitions and in 1917 produced an important paper on screw gauges. (See also Appendix B)[37]

Crompton recalled how, from 1904 onwards, he became increasingly preoccupied with the question of standardization in electrical plant. In August that year, the Institution of Electrical Engineers asked him to accompany their President, J.K.Gray to the United States to represent British electrical engineering at the International Exposition at St.Louis.[38]

The American Institute of Electrical Engineers (IEEE) had set up a committee on electrical standardardization some seven years previously, and the paper on Standardization of dynamo-electric machinery and apparatus, read by Crompton at the St.Louis Congress was followed by a lengthy discussion. Even at this juncture, the Americans were alive to the danger that fixed standards might block technological progress. It was generally agreed, however, that performance as opposed to design should form the basis of electrical standards.

In the discussion he emphasised that electrical standardization should be an ongoing process:

'It appears desirable that standardizing, based on the opinion of the majority of engineers at any one time, is liable to be and should be corrected from time to time as our knowledge increases. With such an understanding standardization may go on simultaneously in all countries, and I hope that those countries who wish to take the matter up will look on it in this light.'[39]

8.10 International Electrotechnical Commission

The Resolution approved for the setting up of an international commission for international electrical standardization by the Chamber of Government Delegates at the St.Louis Congress in 1904, and which followed directly from Crompton's paper on standardization, has been reproduced in Chapter 6.

The history of the International Electrotechnical Commission (IEC) is outside the scope of this book, but since the action taken by the Institution of Civil Engineers (ICE) and the Institution of Electrical Engineers (IEE) leading up to the foundation of the IEC is directly related to Crompton it is summarised below.

On his return from the United States, Crompton proposed to the ICE Council in November 1904 that the St.Louis Resolution should be referred to the British Engineering Standards Committee for consideration and report. Crompton's proposal was for the setting up of a Committee to deal exclusively with International Electrical Standardization, which Committee, if appointed, would correspond with the Chief Electrical Society in America, and would arrange with other countries for the formation of a permanent International Committee on Electrical Standardization.[40]

In February 1905 it was decided that the President of the ICE should confer with the President of the IEE (Mr A.Siemens). This was relatively simple since the 'Electricals' before moving to their present address at Savoy Place were for a number of years located at the 'Civils'.

It was decided by the Council of the ICE '...that the appointment of such a Commission though in every way desirable, would at present be premature; but we believe that preliminary action may, with advantage be taken with the aim of paving the way to the ultimate formation of such a Commission, if the Council approve the general object...'

In December that year, the Secretary of the ICE wrote to the IEE Council – the letter had been drafted by Crompton – that the Engineering Standards Committee (ESC) was 'in favour of the formation of a permanent International Commission on the standardization of the nomenclature [see Appendix D] and ratings of electrical apparatus and machinery', and requested names of suitable persons to serve on an Executive Committee.[41]

On the IEE side, meanwhile, on December 7, Crompton, as a Past-President, reported to the IEE Council that favorable replies had been received from nine countries to the setting up of the Commission. On the President's proposal, the IEE Council then agreed to form this Executive Committee with Crompton acting as Honorary Secretary.

On June 21, 1906 Crompton on behalf of the Executive Committee reported to the IEE Council that the Inaugural Meeting of the Commission would take place on June 26 at the Hotel Cecil in London. After several amendments had been made to the Draft Rules, the new body, 'the name of which was altered to International Electro-technical Commission' was inaugurated at the adjourned meeting the following day, when Lord Kelvin was unanimously elected by the 14 member countries as President, and Colonel Crompton as Honorary Secretary. (Mr Charles le Maistre was subsequently appointed [General] Secretary.) It was agreed that during the first year the IEE would give financial assistance to the Commission.[42]

The last time that Kelvin and Crompton were to meet was on the occasion of the latter's paper, 'Modern motor vehicles', which he read at the Institution of Civil Engineers in February 1907.[43]

8.11 Crompton's Legacy

Under Crompton's leadership, the IEC grew both in size and influence. In 1926 at the Plenary Meeting of the IEC held in New York, in tracing its history over the first twenty years, he said 'We were the first League of Nations, and we are still that'.[44]

As founder of the IEC, Crompton's experience was called upon that year in drafting the procedural rules of the future ISA (International Federation of National Standardizing Associations) – precursor of the post-war ISO (International Organization for Standardization), and sister organization of the IEC – which was formally inaugurated at Prague in 1928.[45]

Elected Honorary President of the IEC in 1931, Crompton continued to take an active interest in the activities of the Commission up to the age of 86, at which time he felt obliged to give up some of his work.

When Crompton had first undertaken the manufacture of electric lighting apparatus, as he put it 'we had no knowledge of the laws governing electrical phenomena, for the names that we now use – Volt, Ampere, Watt – were not introduced until two years later at the Paris Exhibition of 1881.'

In his *Reminiscences* published in 1926, in a tribute to Lord Kelvin's lifelong influence on his work, Crompton wrote:

> 'It was a happy day for us when we were carrying out our order for the Glasgow and South-Western Railway Company to light St.Enoch's Station in Glasgow with arc lamps, that a gentleman came up to me and introduced himself as Thomson, Sir William Thomson, afterwards Lord Kelvin. He there and then discussed with us the form that our arc lanterns must take to throw the least possible shadow on to the platform below them. From this time onwards [1880], right up to the time of his death, I had the benefit of his great scientific knowledge and of his advice.' [46]

Notes

(1) *Certificate of Candidature for Fellowship of the Royal Society.*

(2) Crompton – The Engineer, *Electrical Review*, 23.2.1940, p 210

(3) R.E.Crompton and G.Kapp, On some new instruments for indicating current and electro-motive force, *Journal of the Soc. of Telegraphic-Engineers and Electricians*, Vol XIII,1884, pp 74–84

(4) Brian Bowers, *R.E.B.Crompton, Pioneer Electrical Engineer*, Science Museum, HMSO,London, 1969

(5) *R.S.Catalogue*, 1.5.1895, p 7

(6) R.E.Crompton, On the progress of the electric light, *Journal United Services Institution*, Vol 25, 1882.
NOTE
Zénobe Théophile Gramme (1826–1901) entered the Alliance dynamo and arc lamp factory in Paris as a workman. In 1870, in reinventing the armature, he made it possible for an efficient dynamo to be designed. In the absence of sufficiently developed design criteria, Gramme succeeded in designing a dynamo with a fairly well-proportioned magnetic circuit. Gramme, like Crompton, had an intuitive sense of knowing whether a machine was well designed. *Nature*, 31.1.1901, p 328

(7) R.E.Crompton, *Reminiscences*, London , 1928, pp 81,82

(8) *L'Electricien*, 16.4.1887, p 261

(9) *The Times*,21.9.1931
An initial installation was carried out in the Ball Room and Supper Room at Buckingham Palace in 1883 and 1884, after which electric lighting was gradually extended over the rest of the Palace (R.A. letter 8.9.1978 to Author).

(10) Letter 27.5.1885, C 187, Cambridge University.

(11) Siemens: Notes on the metrical system: Discussion, *Journal IEE*, p 307

(12) ibid, pp 278–326

(13) *The Engineer*,13.2.1903

(14) *Reminiscences*, op cit, pp 98,99
L'Electricien, op cit, 1.8.1881, p 361

(15) The Editor of the *L'Electricien* wrote, 'Nous avons ici trop de noms pour exprimer une seule et même chose, d'autres fois au contraire, nous n'en avons pas un seul'. French engineers could not find equivalent terms for newly coined English words. Problems were also experienced when the appropriate symbol had to be chosen from as many as ten alternatives.
L'Electricien, op cit, 15.7.1881, pp 312,313

(16) *Reminiscences*, op cit, pp 107,108

(17) Letter 20.3.1883, C 184, Cambridge University

(18) Draft letter 21.3.1883. LB4.151, Cambridge University

(19) Letter 4.11.1886, C 188, Kelvin Correspondence, Cambridge University:
Reminiscences, op cit, p 113

(20) The name 'Dynamicals' was adopted at the third meeting of the Society which undoubtedly influenced the development of the electrical industry in Britain. Important papers were read and discussed at their meetings. Apparently, some members – probably after wining and dining – had felt that 'ohm-sweet-ohm' would have been a more appropriate name. *The Electrician*, 29.6.1923; Stanley Steward, The Dynamicals: 95 years of amicable dining, *Electrical Review*, 20.1.1978

(21) *L'Electricien*, 21.2.1885

(22) *Reminiscences*, op cit, p 146

(23) *The Mikado* [1885], Act 1

(24) Crompton Collection, Letter A 21–3, Science Museum, Kensington. Neil Sutton's edited version of 'My little eye' appeared in *Electrical Review*, 26.9.1980.

(25) R.E.Crompton, Presidential Address to IEE, *The Electrician*, 18.1.1895, pp 324,325. R.E.C. Crompton, The cost of the generation and distribution of electrical energy, *Min. Proc. ICE*,106,2

(26) *The Electrician*, 25.1.1895

(27) *Proc.IEE*, 1885, pp 510,511

(28) *Reminiscences*, op cit, p 140

(29) Letter 4.6.1889, *Life*, op cit, p 887

(30) Letter to R.S. 22.11.1911
NOTE
Following on the work begun with the setting up in 1875 of the International Bureau of Weights and Measures in France, which for over a century has ensured agreement between the measurement standards of the participating countries, Germany was the first country to recognize in 1887 the need for accessible standards of the prime quantities of measurement (Physikalische Technische Reichsanstalt), the second was the Russian Mendeleeve Institute (VNIM) in 1893, the United Kingdom's NPL in 1900, and the National Bureau of Standards in the United States in 1901. H.E.Barnett, The British Calibration Service, *Electronics and Power*, Feb.1971, p 64

(31) *The Times*,20.3.1902

(32) *Life*, op cit, p 1165

(33) Personal communications Director of NPL, 2 and 18.8.1971

(34) Draft letter, 13.2.1905, LB 25.26,Cambridge University

(35) R.E.B.Crompton, Unresolved problems in electrical engineering, read on 10.4.1905 – James Forrest Lecture 1905, *Proc. ICE*, Vol cl xii, 1905
NOTE
Sir John Wolfe Barry, Past-President of the ICE, in complimenting Crompton on his lecture, reflected on how much had been achieved within the short history of electrical development:
'Unsolved problems would remain unsolved until they were attacked by men like the lecturer, and those who were working in the same fields...What the end of the twentieth century would be it was impossible to foresee, but one thing was certain, namely, the intimate relation that existed between the different branches of engineering. It was a great pleasure to those who, like himself,were occupied with constructional work, to realize how intimately that and the mechanical and electrical lines of thought were interdependent, and how much one must be the handmaid of the others. It seemed as if one could not exist without the others, and that it was to the development of all that must be looked for that great future which would add so much to the pleasure of life, and would solve so many social problems which were pressing upon the rising generation.'

(36) In 1931 it became the British Standards Institution (BSI)

(37) *Reminscences*, op cit, pp 141,142
R.E.Crompton,Notes on screw gauges, *IAE Proc.*,1917, p176

(38) *Reminiscences*, op cit, pp 205,206

(39) Col.R.E.B.Crompton,Standardization of dynamo-electric machinery and apparatus, *Transactions of the International Electrical Congress*, St.Louis,1904, pp 768- 801

(40) Minutes of the Main Committee (Engineering Standards Committee), Institution of Civil Engineers, 8.12.1904.

(41) MS ICE Council Minutes, 1905, pp 318,370,384–386,539,561

(42) MS IEE Council Minutes, 1905: pp 30,95–97; 1906: pp 153,162–167 *Journal of the IEE*, Vol 39,1907, p 757

(43) *Life*, op cit, p 1196

(44) *IEC Report of New York Plenary Meeting*, April 1926

(45) NOTE
Writing to his family on February 11, 1926, at the age of 81, he recounted that over the previous two days: ...I went to Le Maistre's office [Le Maistre was the first General Secretary of the IEC] and met Glazebroook and Pendred, the Editor of the *Engineer*, and we spent two or three hours very carefully devising the rules which must govern the procedure of an International Mechanical Standardization committee, in order to do for the mechanical and general world that requires standardization what we in the past have done for the electrical world. Le Maistre had prepared a most elaborate rules of procedure which we had to criticize and simplify... He had attended a 'a very difficult committee of the leaders in the Standardizing movement to devise means by which the use of the mark fixed to apparatus to show that it is in accordance with standard specifications can be usefully applied...' He had lunched at the Savoy with the Municipal authorities 'to protest against Winston's raiding the Road funds'. He had participated in a Council meeting of the ICE – and subsequently lunched with the President 'to discuss the constitution of the Council of the Civils to make it more representative ... they have got themselves into such a senile condition' – he explained to Sir William Ellis 'how the Electricals did it, in their case having a supply of younger men on the Council who could put them wise on what the younger men are thinking'. He gave evidence as a witness before the Electricity Commissioners 'to allow poor people to have electric light'. On 11 February, he went to the Ministry of Transport to discuss transport facilities in the event of a general strike. After writing up a record of all these meetings, that evening Crompton attended a banquet of the IEE to receive the Institution's highest award, the Faraday Medal.
Letter 10.2.1926,A21–57 and Personal Diary,1926,Crompton Collection, Science Museum.

(46) R.E.Crompton,Development of Electric Lighting, *Practical Electric Engineering*, c 1920, pp 393–403. *Reminiscences*, op cit, p 88

James Watt and the metric system

Extracts of letter from Prof Archibald Barr , Director of the James Watt Engineering Laboratory, University of Glasgow to Lord Kelvin.*

20th February 1904

Dear Lord Kelvin,

I am much interested in the present movement for the adoption of the Metric system of weights and measures, and have been reading the pamphlet giving extracts from your evidence in the United States with great pleasure.

I suppose you are aware that James Watt took a very active part in the earliest movement towards having a universal system. I have not been able to find whether he had much influence on the French philosophers, but his letters written in 1783 contain some very good passages. Perhaps you will not object to me calling your attention to them − no doubt you know of the existence of these letters − in view of the coming controversy. Here is one passage in a letter to Mr. Kirwan − Nov.14, 1783 − Muirhead's *Life of Watt*, Vol 2, p 180:

> 'Having lately been making some calculations from Messrs. Lavoisier and De la Place's experiments and comparing them with yours, I had a great deal of trouble in reading weights and measures to speak the same language; and many of the German experiments become still more difficult from their using different weights and different divisions of them in different parts of that empire. It is therefore a very desirable thing to have these difficulties removed, and to get all philosophers to use pounds divided in the same manner, and I flatter myself that may be accomplished if you, Dr Priestley, and a few of the French experimenters will agree to it, for the utility is so evident, that every thinking person must immediately be convinced of it.
>
> > [Barr's] NOTE: This proposal to confer with French philosophers was made by Watt 7 years before the French Government invited the Brit Gov. to join in a movement for a common standard.
>
> My proposal is briefly this; let the
> > Philosophical pound consist of 10 oz or 10,000 grains
> > > the ounce of 10 drachmas or 1,000 grains
> > > the drachma 1,00 grains

Let all elastic fluids be measured by the ounce measure of water, by which the valuation of different cubic inches will be avoided, and the common decimal table of specific gravities will immediately give the weights of those elastic fluids.

Dr Priestley has agreed to this proposal and has referred it to you to fix upon the pound if you otherwise approve of it.

I have some hopes that the foot may be fixed by the pendulum, and a measure of water and a pound derived from that, but in the interim let us at least assume a proper division, which from the nature of it must be intelligible as long as decimal arithmetic is used.'

Again on 23rd Nov. 1783 Watt wrote to [Jean André] De Luc:**

'......I proposed to Dr Priestley and Mr Kirwan to agree on a perpetual decimal subdivision of the pound.

"All elastic fluids to be measured by the ounce or pound measure. The decimal tables of specific gravities will give the weight without calculation. All liquids to be weighed......I will be obliged to you to write to M.De la Place on the subject".

[NOTE.- Ten lines of Prof Barr's letter omitted here.]

I wish that later engineers had so much good sense in this matter as Watt seems to have had. It would have been much easier to have made the change in his day that it will be now but I sincerely trust that the offer of your Lordship and those who are acting with you will be successful. I hope too that a comparatively short time will be given for the transition. I think two years will be quite enough time – more especially two years – if unfortuately we have them – of dull trade in engineering.

No doubt a good deal will be made of the cost to manufacturers and merchants. Manufacturers will however as you have well said, be amply repaid and I believe even retail shopkeepers will be well repaid for the outlay on new weights (I mean pieces of metal) and meter rules.

I trust your Lordship will excuse so long a letter on this subject, and again wishing this movement all success.

*   Kelvin Collection, Cambridge University Library, DS4
**   See also Paul Tunbridge, Jean André De Luc, *Notes and Records of the Royal Society London*, Vol 26, No1,June 1971, pp 15–33

Kelvin and screwthread standards

The important issue of standardization of screwthreads – where the need for fastening devices in all types of apparatus or machinery has always been self-evident – was if anything even slower in Great Britain and the United States than was the case with the recognition of the metric system. Essentially the question of international screwthread standards could only be completely resolved once the metric system had been accepted on a worldwide basis.

In Britain, in the latter part of the 19th century, although Whitworth's standards had for many years been generally applied for the larger screwthreads used in heavy machinery, no uniform system existed for the more delicate electrical apparatus now being manufactured in quantity numbers by the growing electrical industry. Each manufacturer had his own system of screwthreads with various diameters, pitch of thread, and number of threads per inch, or per millimetre.*

In 1882, despite his numerous preoccupations as professor at Glasgow University and his membership of many committees as well as his business activities, the question had assumed sufficient gravity for Sir William Thomson to be asked to sit as a regular member on a British Association Committee for the purpose of determining a system of small screwthreads for telegraphic and electrical applications. The distinguished members of the Committee also included Sir Joseph Whitworth, himself, R.E.Crompton and Dr C.W.Siemens.

The Committee's recommendations were published in the annual *B.A. Reports*, and in 1884 it proposed a screw gauge which was too complicated for general adoption. The following year, the British Association reported that the standard series of screws – which had originated in Switzerland where it had obtained the support of leading manufacturers – had been officially adopted by the Telegraph Department of the Post Office and it was hoped that by this fact this would ensure their general adoption by the electrical industry.

The *B.A.Reports* for 1882 and 1884 refer to the Socièté des Arts de Genève that in 1876 had set up a committee for the uniformity of screws. The work of this committee was communicated in several reports by a brilliant but relatively unknown Genevese scientist, Prof Marc Thury. Thury was highly commended in both the above *B.A.Reports* to the effect that he 'had done for the small screws used by watch, clock, and scientific instrument makers for what was done forty years ago by Sir J.Whitworth for the larger screws used by engineers'.**

In the *B.A.Report* for the meeting held in Glasgow in 1901, with Lord Kelvin as Chairman, it was reported by Prof Dalby that the committee had now been sitting for about twenty years endeavouring to standardize small screwthreads. They had managed to produce a universal standard with a simpler form of thread but as this had still not been accepted in Paris, it remained in the form of a recommendation.

The lack of progress in screwthread standardization continued to act as a brake on foreign trade and commerce. It required two major World Wars, with all the complications caused by differing national screwthreads and fastening systems between the allied forces, before the question was eventually considered inportant enough to be dealt with at the international level by the International Organization for Standardization (ISO).

---

*   Sir Joseph Whitworth's classical paper *An uniform system of screwthreads* had been read at the Institution of Civil Engineers, London, in 1841.

**   Prof Marc Thury, *Systématique des vis horlogères*,Geneva, 1878 ; Notice sur le système des vis de la filière suisse,Geneva, 1880 (For an example of Thury's scientific ability see Paul Tunbridge: The first experimental air- cushion machine, *Transactions*, Newcomen Society, Vol 51,1979–80,pp 41–56).

Appendix C
Metrication in Britain

In 1965 the Government decided that metric units should be adopted but on a sectorial basis, industry by industry, so that in time the metric system would be the primary system of weights and measures in Britain.

In announcing the adoption of the metric system to the House of Commons, the President of the Board of Trade pointed out that standards, which were closely linked with metrication, and wherever possible internationally recognised, would be based on the metric system. In order to promote the metrication process, official tenders for procurement issued by the Government and public authorities should be drafted in terms of metric specifications.

It was not considered possible for the changeover to be made by a certain date limit since the problems involved were too complex. It was therefore decided that metrication should be phased over a number of years to enable individual firms to decide for themselves when the time was ripe for metrication change in line with their new product design and the consequent replacement of obsolete machinery and equipment.

The role of the British Metrication Board, which has no statutory or executive powers, was therefore primarily to co-ordinate all the different aspects of this voluntary metrication both in industrial and non-industrial sectors. While it would not be expected to make decisions on the units to be adopted, it would ensure the overall coherence of the metrication process by means of consultation, advice and information of the various sectors involved.

Seven Steering Committees were set up to oversee the work in different sectors, mainly through Trade Associations and similar bodies:

- Agriculture, forestry, fishery and land distribution,
- Food and consumer goods,
- Education and industrial training,
- Engineering,
- Fuel and power,
- Industrial materials and construction,
- Transport and communications.*

In spite of this encouragement given by the Government, including the training of staff in the use of SI units, the transition to metrication has in certain sectors proved disappointing. The United States, with its large domestic markets and smaller international commitments, has been even less inclined to promote metrication, which in turn has influenced certain British manufacturers who export to the USA.

Even in 1990/1991, the correspondence columns of British newspapers such as *The Times* and *The Daily Telegraph* featured a long exchange of letters from readers covering various professions many of whom remain opposed to metrication.

A recent report, 'The failure of British Metrication', sponsored by the Institution of Electrical Engineers, traces the development of metrication over the last 25 years. In contrast to Japan where the changeover has been completed, Britain and the United States remain out of step with the rest of the world and Europe.

Under the heading, 'Stagnation on going metric costing £5bn', the *Daily Telegraph* summarized the findings of the report which concludes that Britain is only inching its way toward the metric system. European Community directives call for metrication to be completed by 1999 but since the Metrication Board was dissolved in 1980, according to the report, 'the metric conversion process in the United Kingdom did not fail completely, but it did go horribly wrong.'**

---

*   F.E.Parr,The Metrication Board: Its Function. *Standardization*, Directorate of
     Standardization, Ministry of Defence, January 1972, pp 22,23
**   *The Daily Telegraph*, 16 Oct 1991

International electrotechnical terminology*

Shortly after Kelvin's death in 1907, a long report by Crompton on the new international electrical commission and standards appeared in *The Times*.

It announced that local or national committees had been, or were about to be, set up in the following countries: Austria, Belgium, Denmark, England, France, Germany, Hungary, Mexico, Spain, Sweden and the United States as well as Australia, Canada, Italy, Japan, New Zealand, Russia, South Africa and Switzerland.

In line with the aims of the Commission, which was to promote 'the adoption in all countries of standard words, expressions, and definitions to denote the machinery, apparatus, instruments, laws, and formulae which are in everday use in electrical engineering', said Crompton, the English Committee – which was the first established – had done much to fulfil this resolution.

Its first task had been the appointment of a committee under the chairmanship of Mr A.P.Trotter, Electrical Adviser to the Board of Trade, [1899–1917], to consider the terms and definitions employed in the electrical profession, 'many of these definitions being at present far too loosely expressed, and therefore requiring revision and standardization'.**

In the interest of simplification, the languages at this juncture were restricted to French and English, but the other countries under the authority of the Commission would draft and publish authorized translations in their languages. It was only after the glossary of terms and definitions had been settled that the more difficult task of the standardization of machinery would be undertaken.

Crompton referred to the recent death of Lord Kelvin, whose 'position had been so pre-eminent in the international world of science that his election as the first president was a matter of course'. After mentioning that the name of Kelvin had been proposed for the Board of Trade Unit of electrical energy [this proposal, referred to in Chapter 7, which had emanated from A.P.Trotter, was not in fact adopted], Crompton wrote 'we realise how profoundly the weight of Lord Kelvin's great name has added to the prestige of the International Commission and lightened the labours of those who have carried out the initiatory work'.

Today, the International Electrotechnical Commission (IEC), is the authority for world standards in electrical and electronic engineering. One important function of the 41 National Committees of the IEC, represented in more than 200 specialised committees, is the compiling of the *International Electrotechnical Vocabulary* (*IEV*) containing over 100,000 terms in nine languages with definitions in French, English and (since 1970) Russian. The *IEV* forms the basis for international communication in electrotechnical fields and as such, is an invaluable tool for science, trade and industry.

The constant evolution of electrotechnical terminology calls for the introduction of new terms or the elimination of outdated nomenclature: this is an essential task of the International Electrotechnical Commission in keeping abreast of developments taking place in industry. The numerous published chapters of the *IEV* cover such varied sectors as general electronics, machines and transformers, recording and reproduction of sound and video, radio communications, physics and chemistry.

International agreement on terminology is essential to enable engineers and scientists and business executives the world over to use a common language in their technical discussions, contracts and specifications.

Scientific and technical language continues to evolve. Evolution has often occurred on parallel lines in different countries, and each scientist has coined his or her own terms. A scientist's colleague in another country often translates these into his or her own first language and they are adopted by a general public without their precise meaning always being understood. Thus we sometimes find cases in which two completely different terms are used to designate the same concept. In such fortunately rare instances a heavy responsibility rests upon the engineer or translator concerned, who in weighting the scales in favour of one or the other term promotes its adoption internationally. What can cause real chaos is if the same term is inadvertently used in different places for different or slightly different concepts.

Faced with an unprecedented explosion in technological progress in the electrical, electronic and telecommunications sectors, whose impact on our daily lives cannot yet be estimated, where each day sees some new development, the IEC infrastructure produces standards which are essential in ensuring coherence between the interdependent machines, installations, and apparatus manufactured by the electrotechnical industry. But IEC standards can only be prepared and effectively put to use if they are universally understood and interpreted in the same way. This is the key role of the *International Electrotechnical Vocabulary* at a time when international trade is assuming even greater importance.

The aim of the *Vocabulary* is to provide precise, brief, correct definitions of internationally accepted concepts and then to provide terms by which these defined concepts are to be known. Those who undertake the work have to first identify such well-defined concepts in different modes of thinking and at times in languages which do not have a common origin and then assign terms that are consistent with the spirit of each of the languages concerned. For the choice of terms due consideration is given to the *Principles and methods of terminology* (ISO 704), published by IEC's sister organization the ISO.

International agreement on scientific terms and definitions ensures that IEC vocabulary remains abreast of developments in science and technology.

But as Prof. Hamburger has pointed out, it is far easier to introduce a new term for an invention or a discovery than it is to fight against an obviously incorrect one which has already been generally adopted.†

Published chapters of the IEV are periodically reviewed and updated where necessary after comparison with other newly issued specialised chapters. The compilation of a general index has enabled duplicated terms to be eliminated; sometimes, however, a term is defined in several ways according to the specialised branch in which it is employed. The *IEC Multilingual Dictionary of Electricity* supplements the IEV.

In all this activity, the IEC works in close cooperation with many other international organizations including in particular the International Organization for Standardization (ISO), the International Telecommunication Union (ITU), the United Nations Economics Commission for Europe and other specialized agencies of the United Nations family and subsidiary bodies.

*   See Author's article, Creating the International Electrotechnical Vocabulary, *Multilingua*, 2–1(1983), pp 35–37

**   A.P.Trotter had also been Editor of *The Electrician*,1890- 1895 (see *Who was Who 1941–1950*, where his recreation is given as 'remembering that he is no longer a Government official'!).

†   Prof Hamburger (Ecole Polytechnique Fédérale, Lausanne) was for many years Chairman of IEC Technical Committee No 1: Terminology.

Appendix E

A letter by William Thomson
on the 'Thomson Effect'*

Of two letters written by William Thomson (Lord Kelvin) to the Genevese physicist Auguste de la Rive (1801–1873) preserved in the public University Library of Geneva, one is of distinct interest. This letter (M.S. 2319), written on 17 December 1856, throws sidelights on the discovery of the 'Thomson Effect' (originally described in his paper to the Royal Society of Edinburgh in 1851) and on the state of his thought about the nature of the mobile element involved in electrical conduction. At the time of receiving the letter de la Rive was still engaged in preparing the third and last volume of his celebrated *Traité de l'électricité*. There is not much doubt that Thomson's letter was occasioned by a passage on thermoelectric effects which had appeared in de la Rive's second volume, published in 1856, the text of which reads (Author's translation):

> 'Clausius comments that thermo-electric effects not only occur at the points of contact of various substances, but that an electric current produces different thermal effects in the same metal, depending on whether it flows from hot to cold or from cold to hot; in other words, in the interior of the same metal where the different portions are at different temperatures, the heat tends to cause the current to flow in a determined direction, as in the case of two metals in contact. It should be added that he clearly considers, as we have done, the manifestation of the electrical state in the interior of the same metal not as a direct effect, but only as a secondary effect of the difference in temperature, i.e., as the result of a change caused by the heat in the molecular state.'

With some justification, Thomson probably felt that his published contributions to thermo-electric theory had been overlooked by de la Rive, as his letter shows:

> 'I take the liberty of asking you to accept a packet of papers chiefly on heat and electricity, which will be delivered to you by a little brother-in-law of mine, Walter Crum, at present living in Geneva. I thought on writing to you before, regarding the convection of heat by electric currents of which you attribute the first notice to M. Clausius, in a slight reference which you make to the subject in your Treatise on Electricity. You will find I believe the very first allusion to the phenomenon of the Electric Convection of Heat in

a paper of mine communicated to the Royal Society of Edinburgh, Dec 15, 1851 (see Proceedings R.S.E. of that date, or Philosophical Magazine, June 1852). Clausius writing about two years later, does not admit the necessity of my inference from observed facts regarding thermoelectric currents, that a thermal effect, reversible with the direction of the current, must be produced by a current through an unequally heated conductor; but he observes that such an effect as this, if it exists, is to be explained by attributing to electricity the property of carrying heat with it when it moves from a hot part to a cold part in an unequally heated conductor.

This, to the best of recollection, is the substance of all that Clausius says on the subject. I do not think he lays claim to any discovery regarding it. It is of course obvious that that peculiar thermal effect which I anticipated may be called a convection of heat by electricity in motion. As early as the month of June or July 1852 I had obtained, in an experiment an indication of a cooling effect on a thermometer in an iron conductor by an electric current of which the nominal direction was from a hot part of the metal to a cold part with the thermometer in an intermediate position. I had a strong tendency from that experiment, to conclude that the true direction of the current is the reverse of the nominal direction, (of the so called 'positive' electricity); and that resinous electricity is the fluid, and 'vitreous electricity' a deficiency of electric fluid – I waited however for a confirmation of the experimental result; which confirmation I did not obtain till Nov 1854; and, was greatly surprised to find, as I did about the same time, that the reverse effect takes place in copper – It is now quite certain that a current of electricity passing through a conducting bar which is kept hot in its middle and cool at its ends, produces a heating effect in the part of the bar by which it nominally enters and a cooling effect in the part by which it nominally leaves, when the bar is of iron or of platinum; and a cooling effect in the first part and a heating effect in the second, when the bar is of copper or of brass.

These results of course leave us as far as ever from fixing on one electricity or the other, as *the fluid*, if electrical phenomena are to be explained by an electric fluid: but they are in perfect accordance with the theory of the mechanical relations of heat and electricity by which I was led to make experiments for the purpose of finding them. From the experimental result for platinum, and from merely thermo-electric experiments which I have made, I can infer with *perfect certainty* that the electric convection of heat in mercury is of the same kind as in platinum and iron; with nearly perfect certainty that it is of the same kind in palladium as in those three; and with high probability that it is also the same in kind, but less in amount, in nickel. These researches form the first part of a rather long paper which is at present passing through the press, and of which I hope to have the pleasure of sending you a copy in the course of a few weeks. The theory of the Electric Convection of Heat is fully given in one of the papers which you now receive; and the results of experiments on the subject up to 1854 (May) are stated Dyn. Theory of Heat 112...114, 121...127.

I enclose a short and very imperfect article which I sent to the Athenaeum some time ago in reply to objections which Mr. Wildman Whitehouse had raised against my theory of the propagation of electricity in submarine telegraph wires. More recently, by conversation with Mr. Whitehouse, by seeing some of his experiments, and by going through, along with him, the results of long and careful investigations which he had made with a view to testing various questions, theoretical and practical, on the subject, I have found all that was wanting for a complete explanation of the apparent discrepancies. Nothing can be more perfect than the agreement of these experimental results, with the theory; and they will afford many very remarkable illustrations of some of the mathematical expressions given in my first paper on the subject, Proceedings Royal Society of London, May 24, 1855.

What I have thus learned from Mr. Whitehouse has also convinced me that the method he has invented for telegraphing through great lengths of submarine wire is likely to be most successful and convenient in all cases, and to give a sufficient rapidity of signalling to make the Atlantic Telegraph a highly advantageous undertaking. It is hoped that a cable, establishing electric communication between Valentia in Ireland and Trinity Bay near St. John's New-foundland (1900 British miles distance) will be laid below the Atlantic before the end of June next – If it is successfully laid, I have no doubt but that it will work successfully.'

* Paul Tunbridge, A letter by William Thomas, FRS, on the 'Thomson Effect', *Notes and Records of the Royal Society London*, Vol 26, No 2, Dec 1971.

Crompton's Patents 1878-1899

PATENT NUMBER	DATE	TITLE
2721/78	8 Jul 1878	Valve – apparatus for controlling motion of fluids
4753/78	22 Nov 1878	Valve – apparatus for controlling motion of fluids
245/79	21 Jan 1879	Electric – lighting
3509/79	2 Sep 1879	Regulating mechanism for electric lamps
4598/79	12 Nov 1879	Casting metal pipes or tubes
69/81	6 Jan 1881	Casting metal pipes or tubes
5080/81	21 Nov 1881	Conducting and distributing electric currents
5159/81	25 Nov 1881	Galvanic batteries and electro-chemical accumulators
346/82	24 Jan 1882	Electric lamps
2618/82	3 Jun 1882	Dynamo-electric machines
2619/82	3 Jun 1882	Electric – lighting
3339/82	14 Jul 1882	Arc regulator lamps
4810/82	10 Oct 1882	Dynamo-electric machines
5438/82	15 Nov 1882	Conduction and distribution of electric currents
67/83	4 Jan 1883	Governors for steam engines, &c
1877/83	13 Apr 1883	Measuring electric currents, &c
2539/83	22 May 1883	Arc regulator lamps
4453/83	18 Sep 1883	Measuring electric currents and electro-motive force
5945/83	31 Dec 1883	Governors
4302/84	3 Mar 1884	Dynamo – electric machines
4794/84	12 Mar 1884	Measuring currents of electricity
8063/84	22 May 1884	Regulating electric arc lamps
11453/84	20 Aug 1884	Magneto-electric and dynamo-electric machines and electric motors
2490/85	24 Feb 1885	Distributing currents of electricity
11987/85	8 Oct 1885	Cutting rack teeth
12019/85	9 Oct 1885	Holophotal projectors
136/86	5 Jan 1886	Electrical tramways or railways
1183/86	23 Jan 1886	Raising, lowering and supporting lanterns, &c
3474/86	11 Mar 1886	Sight drop lubricators
3475/86	11 Mar 1886	Dynamo-electric machines

PATENT NUMBER	DATE	TITLE
3508/86	12 Mar 1886	Electric light fittings
5168/86	14 Apr 1886	Electric lighting of trains
7004/86	25 May 1886	Generating electric currents
12880/86	9 Oct 1886	Armatures of dynamo-electrical machinery, &c
14269/86	5 Nov 1886	Galvanometers
17120/86	31 Dec 1886	Dynamo-electric machinery
275/87	7 Jan 1887	Distribution currents of electricity
2122/87	10 Feb 1887	Electric motors
6409/87	2 May 1887	Dynamo-electric machine and motor
6681/87	6 May 1887	Secondary batteries or electrical accumulators
6754/87	9 May 1887	Dynamo-electric machines
9905/87	14 Jul 1887	Underground electric conductors
9979/87	16 Jul 1887	Circulation of electrolyte in electric depositing
12437/87	14 Sep 1887	Electric lighting
17599/87	22 Dec 1887	Dynamo-electric machine
5811/88	19 Apr 1888	Generating and distributing electrical power
7655/88	25 May 1888	Electric conductors and conduits
8099/88	4 Jun 1888	Electric hoists or cranes
12437A/88	13 Jun 1888	Electrical measuring instruments
13687/88	22 Sep 1888	Electrical furnaces
4901/89	20 Mar 1889	Abandon
11589/89	20 Jul 1889	Electric conductors
12901/89	15 Aug 1889	Secondary batteries
17623/89	5 Nov 1889	Acid resisting coverings
14698/90	17 Sep 1890	Conductors distributing electrical energy
184/91	5 Jan 1891	Armatures of dynamo machines
184/91	5 Jan 1891	Dynamo-machines
8499/91	19 May 1891	Controlling electric currents
486/92	9 Jan 1892	Variable resistances for electrical purposes
487/92	9 Jan 1892	Producing decorative advertising & effects by electricity
1878/92	1 Feb 1892	Protecting electrical conductors
2162/92	4 Feb 1892	Arc lamp
2163/92	4 Feb 1892	Electric measuring apparatus
5094/92	15 Mar 1892	Linings or coverings for walls, ceilings
7252/92	14 Apr 1892	Electrical distributing apparatus
12350/92	4 Jul 1892	Materials for upholstering
12776/92	12 Jul 1892	Regulating temperatures in making metals &c
15840/92	3 Sep 1892	Electric arc lamps
17091/92	24 Sep 1892	Insulating and supporting electric wires
17092/92	24 Sep 1892	Electric heating, cooking &c

PATENT NUMBER	DATE	TITLE
20476/92	12 Nov 1892	Embossed or moulded materials
20768/92	16 Nov 1892	Embossing sheet metal &c
22951/92	13 Dec 1892	Switchboards for distributing electrical energy
257/93	5 Jan 1893	Insulating electric wires for conductors
258/93	5 Jan 1893	Applying electricity for heating liquids &c
259/93	5 Jan 1893	Drying, heating, &c paints, oxides electricity
10015/93	19 May 1893	Utilizing electrical energy for culinary purposes
10569/93	20 May 1893	Regulating temperature of electrical heating, cooking &c
11427/93	10 Jun 1893	Electric accumulators
17197/93	13 Sep 1893	Dynamo-electric machinery
21752/93	14 Nov 1893	Crimping, bending or shaping, holding wire &c
24790/93	23 Dec 1893	Brush-holders for dynamo-electric machines
11452/94	13 Jun 1894	Connectors for electrical conductors
12181/94	23 Jun 1894	Electrically heated apparatus for soldering, brazing &c
16744/94	3 Sep 1894	Applying heat for therapeutic and purposes
21855/94	13 Nov 1894	Trueing up, scouring &c dynamo-electric machine cylinders &c
1318/95	19 Jan 1895	Armatures for dynamo-electric machines
2604/95	6 Feb 1895	Arc lamps
5515/95	15 Mar 1895	Electric measuring instruments
8664/95	1 May 1895	Arc lamps
9388/95	11 May 1895	Arc lamps
10810/95	31 May 1895	Electric tramways
18305/95	1 Oct 1895	Supporting and insulating electric conductors
12251/96	4 Jun 1896	Pulley-blocks
15326/96	10 Jul 1896	Insulating and supporting electric wires
16019/96	20 Jul 1896	Electric regulating switches
22046/96	5 Oct 1896	Electric switches
26267/96	20 Nov 1896	Electric heating apparatus
27446/96	2 Dec 1896	Electrical resistances
28578/96	14 Dec 1896	Bonding electric railway &c rails
6276/97	10 Mar 1897	Arc lamps
11375/97	7 May 1897	Bicycles or Tricycles
11622/97	10 May 1897	Distributing and using electric currents
13164/97	27 May 1897	Dynamo-electric armatures
20385/97	4 Sep 1897	Furnace-flame bridges

PATENT NUMBER	DATE	TITLE
5898/98	10 Mar 1898	Cycle pedals &c
6695/98	19 Mar 1898	Electric regulating apparatus
19000/98	6 Sep 1898	Switching apparatus
24812/98	24 Nov 1898	Electric meters
26610/98	16 Dec 1898	Arc lamps
3849/99	21 Feb 1899	Gearing for pumps &c
4007/99	23 Feb 1899	Electrically controlled switches
4203/99	25 Feb 1899	Cycle &c, frames

Index

Printed in the USA
CPSIA information can be obtained
at www.ICGtesting.com
JSHW011510221024
72173JS00005B/1269